Aztec Medicine, Health, and Nutrition

Aztec Medicine, Health, and Nutrition

Bernard R. Ortiz de Montellano

Rutgers University Press
New Brunswick and London

Copyright © 1990 by Bernard R. Ortiz de Montellano
All Rights Reserved
Manufactured in the United States of America

Library of Congress Cataloging-in-Publication Data

Ortiz de Montellano, B. (Bernard)
    Aztec medicine, health, and nutrition / Bernard R. Ortiz de
Montellano.
        p.   cm.
    Includes bibliographical references.
    ISBN 0–8135–1562–9 (cloth).   ISBN 0–8135–1563–7 (pbk.).
    1. Aztecs—Medicine.   2. Aztecs—Food.   3. Aztecs—Health and
hygiene.   I. Title.
F1219.76.M43078   1990
972'.018—dc20
                                                            89–70142
                                                            CIP

British Cataloging-in-Publication information available

To my wife, Ana Mercedes Ortiz de Montellano, and my mother,
Thelma Ortiz de Montellano, who sustained me through it all,
and to Charles Dibble who got me started

Contents

Illustrations

Tables

Acknowledgments

I owe much to many people. Charles Dibble has long been a model
and an inspiration. He read and critiqued all of the manuscript,
offering many useful suggestions and supplying materials from
his extensive collection. I consider Angel Ma. Garibay (deceased)
my intellectual grandfather. His pioneering works were models of
wide-ranging scholarship, and his deep knowledge of Aztec culture brought his Nahuatl translations closer to the true Aztec
spirit than did later versions based primarily on linguistic
principles.

My Mexican colleagues, Carlos Viesca Treviño, Xavier Lozoya,
Doris Heyden, and Carmen Aguilera, encouraged, read, and critiqued my work, often providing essential articles, books, and
materials. From Robert Bye and Edelmira Linares I received hospitality, color slides, and ethnobotanical expertise. Miguel and
Chonita León Portilla afforded hospitality, wise counsel, access to
his excellent library, and publications from the Instituto de Historia. Luis and Leticia Vargas gave me a place to stay in Mexico City
and introduced me to a number of other Aztec scholars. Luis was
very helpful in obtaining publications from the Instituto de
Antropología. Alfredo López Austin, whose work has set a high
standard for accurate and comprehensive scholarship, was most

generous with advice, suggestions, and commentary. My intellectual debt to him is evident in the numerous citations of his work.

In the United States, Guido Majno searched the Hippocratic corpus for me, and Joseph Bastien supplied information about folk medicine in the Andes. My co-workers, Carole Browner, Arthur Rubel, Michael Logan, and Robert Trotter, were a constant source of intellectual stimulation and encouragement. Leon Abrams always wanted me to write about Aztec food and read and critiqued the parts on diet and nutrition. Mike and Sophie Coe were gracious hosts and provided me with the results of their research. Kaja Finkler served as a sounding board for my theories and furnished much data from her own studies. Eloy Rodriguez and I have had many fruitful and exciting conversations about ethnobotany, natural products, chemistry, and life in general. Diana Marínez for years has been a source of friendship, jokes, reference materials, and recipes. Karen Reeds, my editor, and Norman Rudnick, my copy editor, at Rutgers University Press, have done yeoman work in helping to shape, clarify, and refine my writing. They significantly improved the organization of the book.

To my mother, Thelma Ortiz de Montellano, I owe a lifetime of support and close reading of my work. My wife, Ana Mercedes, has read and edited everything I've written, including this book. Her support and dedication were essential to the completion of this work. I will always be grateful to her.

This book owes much to many other people, but its errors are mine.

Detroit, Michigan
April 1990

Note on Text

A few general remarks are in order about terms used in the book. The designation Aztec is applied to peoples living in the Valley of Mexico who have a common history and are heir to the culture of Mesoamerica. The term Mexica refers to those Aztecs who resided in Tenochtitlan, now Mexico City, and who ruled the Aztec empire at the time of the Spanish Conquest. Nahuatl is the language in the Utoaztecan language group spoken by the Aztecs. Words in this language are accented on the next to last syllable, and Nahua words, including place names, are so treated, rather than according to Spanish convention.

Several primary sources are cited extensively and discussed fully in chapter 1. Since they were originally written in the sixteenth century but cited from modern editions, a brief description might be useful. The *Badianus Codex* was written in 1552 in Latin, but references are to a Spanish translation published in 1964. Ruiz de Alarcón was composed in Spanish in 1629; references cite either a 1953 Spanish edition or one of two English translations, published in the 1980s, which are the same book although they have different titles. Hernández' Latin work was finished in 1577, but we cite either a partial (1942–1946) or full (1959) Spanish translation. Sahagún's book went through several stages and languages.

Two versions (*Primeros Memoriales* in 1560 and *Códices Matritenses* in 1565) were written in Nahuatl, never fully translated, and printed in facsimile in 1906. A subsequent Nahuatl version, the *Florentine Codex* (1577), was translated into English and published in 1950–1969. Sahagún's Spanish version, the *Historia Natural*, was first published in 1793; a 1956 edition is cited here.

Unless specifically noted, all translations of cited passages originally in Spanish are the author's own.

Aztec Medicine, Health, and Nutrition

Introduction

The ease with which a few Spaniards were able to conquer the Aztec Empire has always presented questions of interest to historians and anthropologists. How healthy were the Aztecs before the Spanish arrived? Can Aztec vulnerability to European disease and abuse be explained by general weakness before the Conquest? In the answers to these questions—and we will argue that, in the main, the Aztecs were a healthy people—Aztec medicine becomes an important part of the issue. The Aztec system of defining health, of explaining the causes of sickness, of diagnosing, preventing, and curing disease is interesting in itself. It acquires further meaning in juxtaposition to European medicine in the sixteenth century and as an extended example of a non-Western medical system. The Conquest meant that both the Spanish and the Aztec encountered an alien set of medical beliefs, traditions, illnesses, and remedies, and had either to destroy or assimilate the other. Each found some aspects of the alien beliefs and customs apparently familiar and easy to respect, others wholly incomprehensible and abhorrent.

The Spanish also confronted and destroyed other thriving civilizations—the Mayans, the Incas—in their expansion in the Americas. We concentrate on the Aztecs because they epitomize a

long tradition of culture in Mesoamerica, because more information is available about their religion and medicine than about any
other American culture, and because the survival of Aztec beliefs
in modern Mexico gives us a chance to cross-check our sources.
Also, the Aztecs and their medicine are of particular interest to
historians, anthropologists, and medical historians because the
Aztec was one of a few state-level civilizations (China, India, Islam,
and Inca are others) encountered by Europeans in the Age of
Conquest.

The existence of writing in state-level societies allows knowledge to accumulate, be passed on, and achieve an increasing level
of sophistication. The Aztecs are a crucial example because written records make theirs the best documented state-level civilization of the New World. Their medicine was ideologically coherent
and sophisticated, with obvious appeal for anthropologists and
medical and religious historians concerned with the nature of
non-Western thought systems. We distinguish between medicine,
the definition and treatment of diseases, and health, a state of
being that depends on the nutritional condition of the population
and its exposure to disease-causing factors. The Aztecs are also
relevant to modern concerns because they erected and maintained a state-level civilization with a high population density in a
not always congenial ecological zone, subject to frost and low
rainfall and lacking domesticable herbivores such as the cow and
horse, as well as the omnivorous pig. The agricultural systems and
food plants that allowed the Aztecs to keep their population
healthy and well nourished under such circumstances could be
used today to help fill the food needs of Third World nations.

The book briefly summarizes Aztec history, culture, and religion, as well as the Spanish Conquest and colonial organization
(chap. 1). These provide a context for the consideration of Aztec
medicine and the process of syncretism, a fusion of Aztec and European influences, that began with the Conquest and changed
both Aztec and Spanish culture in Mexico. Syncretism attaches
new meanings or functions to old elements of a culture, or causes
old elements to be attributed to newly introduced cultural factors.
Whenever two cultures are in contact, syncretism occurs. Also, the
subordinate culture (Aztec) incorporates more elements of the
dominant (Spanish) culture than the dominant culture borrows
from the subordinate one. Beliefs previously similar in both cultures are syncretized more easily.

The alterations due to syncretism require that primary

sources of information be carefully scrutinized and evaluated to obtain an accurate picture of Aztec medicine before the Conquest (chap. 1). Modern Mexican folk medicine is a syncretic mixture of Aztec and Spanish concepts. A clear image of pre-Columbian beliefs, both Aztec and Spanish, is necessary to be able to trace the interplay of these two medicinal traditions from the Conquest to the present. Our review will focus on the concept of "hot" and "cold" in medicine and health, and on a group of "culture-bound syndromes"—*susto, caída de mollera,* and *mal de ojo*—whose origins, native or European, have aroused much interest and controversy (chap. 8).

As the mortality rate of AIDS has demonstrated, the panoply of modern science is useless when the immune system is disabled or otherwise inadequate as the first line of defense. Food and nutritional status are an important component of health, often neglected in discussions of Aztec medicine. However, they are crucial in determining a person's ability to mount an immune response, resist infection, and heal after trauma. The Aztecs had an ample, nutritious, and well-balanced diet, due to their highly productive and labor-intensive agricultural techniques, and to some unusually efficient and nutritious foods (chap. 4).

An aspect related to agriculture is the carrying capacity of the Valley of Mexico, that is, the population the ecosystem could support without incurring irreversible damage. This capacity depends on a complex interaction between the number of people, population density, natural resources, and agricultural systems. Michael Harner (1977) and Marvin Harris (1978) have argued that the Aztec population was excessive and that ritual cannibalism was in fact carried out to obtain an adequate amount of protein. We shall show that this is in error, that the Valley of Mexico had adequate carrying capacity to forestall a population crisis and that a nutritious balanced diet was possible without recourse to human flesh (chap. 3).

Aztec medicine was holistic. Illness and disease were seen as the interaction of supernatural, magical, and natural causes. One factor might predominate in a particular case, but illness causation was multifactorial. We discuss the three causes separately, to illustrate more clearly the process of diagnosis and the relationship between a proposed cause and the recommended cure (chap. 6), but it should be remembered that the Aztecs did not consider them independently.

We shall use the terms disease, ailment, and illness fairly

interchangeably. The distinction in Western biomedicine between disease, a pathological process recognized by a physician, and illness, a cultural construct defined by the patient but not necessarily seen as a disease by the physician, is not very useful in a belief system like that of the Aztecs.

The question of how to evaluate a non-Western medical system fairly has concerned medical anthropologists and medical historians. To assume that only the tenets of Western biomedicine are valid, and that they are the sole basis for judging the validity of different belief systems, is arrogant and unproductive. Western medicine has often rejected effective treatments and accepted useless ones (Goodwin and Goodwin 1984; Moerman 1983). We propose a new methodology that takes advantage of the replicability of biomedicine and of the vast literature on the chemical components of plants, but uses native concepts of disease. First, Aztec medicine is judged according to the biomedically verified production by its proposed remedies of the physiologic effects required by Aztec etiologic beliefs. This is fair because it does not impose Western definitions of disease and cure as the sole standard of judgment. Second, Aztec medicine is also evaluated in terms of its effectiveness according to Western biomedical beliefs and standards. This dual test shows that Aztec medicines were effective, not only by Aztec standards but often by modern standards as well (chap. 7).

Our critical overview of Aztec health and medicine concentrates on public health, nutrition, and medical theory and practice. Guerra (1966), Coury (1969), and Viesca Treviño (1984c) have discussed the social structure of Aztec practitioners and the organization of medical care. We pay special attention to the question that both native and sophisticated observers are apt to raise about any medical system different from their own: Does it really work?

Our approach to evaluating the effectiveness of Aztec medical theory and practice is unusual because it takes the *Aztec* perceptions of diseases, their causes, and appropriate therapies as the starting point. We then ask whether the Aztecs' remedies, usually a holistic mixture of treatments, are likely to have worked given their own explanations of the maladies. For a wide variety of Aztec illnesses, we attempt first to disentangle the complex of magical, divine, and natural forces the Aztecs held responsible, and then draw on contemporary physiology, pharmacology, and psychology to see if the prescribed treatments (whether religious, magical, or rational) could have been empirically effective in counteracting the perceived cause. This technique allows us to judge the effectiveness of Aztec medicinal practice on its own terms without trying to make the illnesses defined by Aztec medicine identical to or congruent with diseases recognized by either sixteenth-century Spanish or twentieth-century American medicine.

Aztec Culture
at the Time
of the Conquest

Historical Background

Pre-Conquest Period

The Aztecs were the last chapter in a very long cultural book.
Their domination over Central Mexico began in the fifteenth century, but for more than two thousand years before them Mesoamerica had been the home of a series of impressive civilizations.
The cultures that spread through this geographic region, extending from roughly the middle of Mexico to El Salvador and Honduras (see map), all shared a number of distinctive features: a
complicated calendar based upon the permutation of a 260-day
sacred calendar and a 365-day solar calendar, hieroglyphic, or picture, writing, a ball game played with a solid rubber ball, chocolate
beans as a monetary unit, a very complex pantheon of deities (Coe
1984: 11–12; Kirchhoff 1943). There existed something like a pan-
Mesoamerican religious system with a common worldview rooted
in the distant past (Caso 1971; Coe 1981). This worldview varied
somewhat in superficial matters between different areas and between different historical periods, but successive cultures over
two millennia shared a set of fundamental beliefs. The Aztecs and
their medical beliefs are the focus of this study because they were

1.   Map of Mesoamerica indicating major regions and important cities. (From G. W. Conrad and A. A. Demarest, *Religion and Empire*. Copyright © 1984 by Cambridge University Press)

the dominant political force at the time of the Conquest and much more information is available about them than about other Mesoamerican cultures, and because their views are representative of those of previous civilizations.

The earliest culture in the Valley of Mexico (see enlarged map section) was derived from the Olmec, as seen in Tlatilco (800 B.C.). The first empire to arise from the Valley was that of Teotihuacan, beginning ca. 100 B.C. Teotihuacan was the first of a series of empires whose power and influence were based on the strategic importance of the Valley of Mexico and the control it afforded over crucial natural resources. Teotihuacan's sway extended from Northern Mexico to the Maya area in Central America and lasted until A.D. 650 when it is believed that the empire succumbed to repeated incursions by bands of hunter-gatherers from the arid regions of what are now northern Mexico and southwestern United States. A repetitive pattern ensued. Nomadic bands assaulted and defeated agricultural empires in Mesoamerica, then settled down in the Valley of Mexico, became farmers themselves, assimilated Mesoamerican culture, and began to build a new empire.

Thus, after a period of chaos following the fall of Teotihuacan, the Toltec empire (A.D. 1100–1250) succeeded in reestablishing

2.   The Valley of Mexico. Lake Texcoco was a salt lake while Lakes
Chalco and Xochimilco were sweet-water lakes. Tenochtitlan, now
Mexico City, was the capital of the Aztec Empire and home of the
Mexica. (From G. W. Conrad and A. A. Demarest, *Religion and Em-
pire*. Copyright © 1984 by Cambridge University Press)

order. Its fall again led to dispersion of the population of the Valley
and its isolation from the rest of Mesoamerica. In the interregnum,
groups of Nahuatl-speaking tribes we call Aztec began to migrate
to the Valley of Mexico. The Mexica were one of the last tribes to
arrive and founded Tenochtitlan in A.D. 1325 as a dependency of
the Tepanecs, who had previously achieved a dominant position
in the area. In A.D. 1428 the Mexica, bolstered by alliance with the
cities of Texcoco and Tacuba, revolted and gained independence.
This Triple Alliance became the basis for the Aztec Empire, which
dominated Central Mexico until the Spanish Conquest (1519–

1521). By the time of the Conquest, Tenochtitlan had become the dominant member of the Alliance.[1]

Aztec civilization developed from a reservoir of Mesoamerican culture in the Valley of Mexico, much of it having withstood periods of fragmentation between successive empires. Its preservation can be attributed to the conservative nature of peasant societies. Oral transmission is the key factor in the selection of the beliefs, ideas, and customs that are faithfully retained. Each time an empire in the Valley of Mexico fell, the state organization and its institutions, such as the religious and educational systems, withered away, but cultural vestiges resided in the people themselves who, though scattered, remained with the land. The situation was analogous to that of Europe after the fall of the Roman Empire. Although the peoples of Europe remained (probably with some decrease in numbers), the structure of coordination and communication maintained under the aegis of Rome disappeared. In Central Mexico, peasants continued to live much as they had under the empire and kept beliefs and traditions alive. However, it is crucial to keep in mind that the people preserved best the things most important to them, including foods, dress, housing patterns, and useful technologies. Among religious beliefs, deities most likely to survive from the abundant pantheon were those most important to peasants: the rain god Tlaloc and other members of the rain-agriculture-fertility complex, for example. Beliefs about deities more closely identified with or worshipped by elites were unlikely to be preserved intact. For example, the Maya calendar, the province of the elite, accumulated errors leading to a difference of two days in the initial day of the year (the year bearer) between the Classic Era (A.D. 200–900) and the time of the Conquest in the Maya zone, and to a one-day shift in the year bearers after A.D. 977 in Central Mexico (Ortiz de Montellano 1979a).

At the time of the Spanish Conquest, the Aztecs had a state-level society, that is, they were socially stratified and had centralized power. They had characteristic state-level institutions such as laws, taxes, a complex administrative organization, a hierarchical institutional religion, police powers, and an army. An exception, which would prove to be important, is their retention of shamanic religious traits more characteristic of hunter-gatherer societies.

Aztec society was stratified into several levels. The top level was composed of *pipiltin*, nobles. The majority of the population was made up of *macehualtin*, commoners, who were organized into *calpulli*, territorial landholding groups with some kinship features whose exact structure is still somewhat controversial. All members of a calpulli claimed descent from a common ancestor and worked lands that belonged to individual families but could not be alienated from the group. Members of the same calpulli had their own schools (*telpochcalli*), temples, and patron deities, and fought together in times of war. Apart from the family, the calpulli was the major structure for social organization in Aztec society (Coe 1984: 154). The *mayeque*, serfs bound to the land, were a third class. The lowest stratum was composed of slaves, *tlatlacotin*. Two groups, long-distance merchants, *pochteca*, and luxury artisans, *tolteca*, occupied ambiguous positions intermediate between nobles and commoners.

Some social mobility existed. A commoner could rise to the rank of noble by virtue of bravery in combat, demonstrated by the number of sacrificial prisoners he captured. Slavery also was neither hereditary nor permanent. One could sell oneself into slavery or be sentenced to it for crimes committed, but one could also win freedom. Children of slaves were free (Berdan 1982: 61–63). Most commoners were educated in the telpochcalli; nobles and priests went to a separate school, the *calmecac*.

Governmental power was embodied in the ruler, the *tlatoani* ("chief speaker"). He was assisted by a Council of Four, which included a vice-ruler, the *cihuacoatl* ("snake woman"), an advisor on military matters, and one who supervised judicial processes. The cihuacoatl, in addition to his governmental role, was the head of the religious hierarchy. Similarly, the tlatoani had numerous religious functions; church and state were intertwined rather than separate. All political matters were supervised by the *tlatocan*, a supreme council composed of 15–20 distinguished nobles who advised the ruler and the Council of Four. The office of the tlatoani was not inherited by primogeniture. He was elected by the council from a fairly limited dynasty of related high-ranking families, chosen on the basis of individual traits of bravery, leadership, and ability. Over time, the status of the tlatoani changed from "first among equals," at the beginning of the empire, to near divinity by 1500.[2]

Aztec religion was an ecclesiastical state institution with a hierarchy of priests ruled by the cihuacoatl, training schools, and state-sponsored rituals. It differed from most state religions in that it retained elements such as shamanism and animism from its past as a hunter-gatherer society. Another distinctive feature was the role of the calendar. A given day was included in both a 260-day and a 365-day calendar. The *tonalamatl* (260-day) combined twenty day names with the numerals one through thirteen. This calendar and its implications for health will be discussed further in chapter 2. The 365-day solar calendar was composed of eighteen "months" of twenty days and one "month" of five days. Each of the long months had a number of scheduled state-sponsored festivals, rituals, and celebrations of patron deities. Rituals were coordinated with agricultural and solar cycles and often designed to request rain and agricultural fertility or to give thanks for harvests. Other ceremonies sought success in war or were aimed at maintaining the universe by "feeding" the sun. Human sacrifice and cannibalism were integral components of Aztec religion. Cosmological beliefs and other religious aspects, particularly relevant to medicine, will be discussed later.

Post-Conquest Period

The conquest of Mexico by Hernán Cortés began in 1519 and was completed by 1521. The Mexica could easily have defeated the small Spanish force from a purely military standpoint, but various factors weakened their defense and contributed to their defeat. The arrival of the Spanish from the east, and in a year called One Reed in the Aztec calendar, seemed to fulfill a myth that the god Quetzalcoatl would return from that direction in such a year. Therefore, the Aztecs at first considered the Spanish to be gods and were psychologically incapable of fighting them. When the less-than-divine behavior of the conquerors became apparent, it was too late. Also, Cortés had the assistance of an Indian slave, Malinche, who served Cortés as a translator and gave him an insight into the culture and psychology of the Aztecs, which he cleverly exploited. Cortés was able to use discontent within the empire to recruit allies such as the Tlaxcalans who wanted to overthrow the Mexica. In the final siege of Tenochtitlan, Cortés was assisted by disease as well as native allies. Smallpox brought by the Spanish devastated the population of the Mexica (MacLachlan and Rodríguez 1980: 68–76; Dibble 1978).

The fall of Tenochtitlan did not destroy established lines of authority within the Aztec Empire, and many Indian groups were willing to accept a substitution of Spanish control for that of the Mexica. As in other colonial situations where the natives greatly outnumbered the rulers, the native ruling class was brought into a governing partnership. Christianity and Spanish law were introduced as instruments of sovereignty and social control. The organizational models used were those developed originally to incorporate Moslems into Spanish society. The municipality functioned as the primary political and administrative unit and in a frontier locale could function as a self-contained governing body. Municipalities were able to perform all the functions of government, judicial, legislative, and executive (MacLachlan and Rodríguez 1980: 77–78).

Urban centers were arranged in a hierarchical manner. The head municipality, or *cabecera*, controlled secondary villages (*villas*), which were surrounded by smaller dependencies (*lugáres*). Important established Indian settlements were incorporated into the structure by designating their chiefs as hereditary rulers and investing them with power as *gobernadores*. These gobernadores were charged with maintaining order, collecting taxes, and enforcing Spanish laws (MacLachlan and Rodríguez 1980: 78).

Tribute that had been paid to the Aztec Empire was appropriated by the Spanish. The *encomienda* placed tribute-paying groups under the authority of an *encomendero* who, in turn, was to protect them. Rather than providing protection, these arrangements were used to obtain a labor force by enslaving the Indians. This action was justified on the grounds that it brought pagans into a close relationship with Christians and facilitated their conversion.

The conquerors disagreed fiercely among themselves about jurisdiction over the Aztecs. In particular, the secular clergy, those who took vows of ordination and were under the authority of bishops, and the regular clergy, members of religious orders who often lived in cloistered communities and took vows of poverty, struggled with one another for power. The two branches of the church fought over such issues as who could convert and baptize the Indians, what powers each could exercise, and who could punish paganism and heresy. When Pope Leo X authorized the first friars (regular clergy) to go to the New World (Franciscans in 1523, Dominicans in 1526, and Augustinians in 1533), he granted them great latitude and powers normally reserved for secular clergy and bishops. Most of the evangelization of the New World was carried out by friars, who were allowed to administer the sacraments and perform other duties normally performed by secular clergy. Not surprisingly, the orders resisted the bishops' efforts to restore secular control over these activities well into the seventeenth century (Ruiz de Alarcón 1984: 3–4).

In the sixteenth century, the struggle over Indian policy centered on the institution of the encomienda. The secular clergy supported the colonists, the encomienda, and the exploitation of the Indians by disparaging Aztec cultural achievements, exaggerating the barbarity of human sacrifice, and exalting the role of the conquerors in bringing relief and salvation to the common people. The more savage and uncivilized the Aztecs appeared, the more justification for the encomienda system and the less the psychological conflict of the colonist with the Christian principle that all men are brothers. Some of the chief proponents of this viewpoint were Francisco López de Gomara, Gonzalo Fernández de Oviedo, and Juan Sepúlveda, influential nobles in Spain (Keen 1971: 77–92).

Members of the regular orders took quite a different attitude toward the Indians. As Phelan (1972) pointed out, the Franciscans, in particular, believed that the Indians were the Ten Lost Tribes of Israel and that their conversion would bring about the Millennium. For them, the Aztecs were rational, intelligent persons whose cultural achievements equaled those of the Greeks and Romans (Keen 1971: 77). The Aztec system of laws, child-rearing practices, education, and rules on sobriety struck them as admirable, far superior to Spanish colonial mores. Although the Devil had led the Aztecs into sin and idolatry prior to the arrival of the Spaniards, once they were converted to Christianity, a great Aztec society could be reconstructed on pre-Columbian organizing principles (Sahagún 1956: vol. 3, 157–168). This attitude toward Aztec institutions led, for example, to Sahagún's ethnographic investigations, which were aimed at preserving information about the knowledge and customs of pre-Columbian society with the hope of restoring worthwhile practices in the future. According to this view, it was unjust to give the Indians away as chattels because they had immortal souls. The encomienda system, especially its abuses, was condemned. The main proponents of this view of the Indians were Bartolomé de las Casas, Bernardino de Sahagún, and Diego Durán (Keen 1971: 92–104, 110–120).

In 1536, following their beliefs, the Franciscans established the College of the Santa Cruz in Tlatelolco, which enrolled children from Aztec nobility for training as scribes and translators of Nahuatl, Spanish, and Latin. The students, in turn, would assist the friars in missionary efforts and in compiling ethnographic documents. This began the process of religious syncretism that led to the folk Catholicism of Mexico today.

Syncretism was facilitated by characteristics of both the Aztec and Catholic religions. Soustelle (1970: 116) commented that Aztec religion was very receptive to the gods of other groups. Every time the empire grew and annexed new provinces, the conquerors brought images of the local gods back to Tenochtitlan's Main Temple to be cared for and worshipped. The great complexity of Aztec religion derives in part from the absorption of deities, myths, and rituals from conquered areas. The Aztecs therefore found it easy to attribute their defeat to the superiority of the God of the Spaniards compared to their own gods and incorporated Him into their pantheon as readily as they had absorbed deities of their own conquests. In Spanish eyes the defeat of the Aztecs was God's punishment for Aztec sins (Sahagún 1986: 113, 161).

In their efforts to convert large numbers of Indians, the missionaries demonstrated a flexible attitude toward native practices and forms of worship. Religious ceremonies before the Conquest had involved large crowds in open spaces and were very lengthy. The "open chapel" architecture of a number of churches, such as Actopan, Otumba, and Coixtlahuaca, is peculiar to New Spain and clearly an adaptation to the Aztec style of worship. It consisted of a large enclosed courtyard with small chapels at its corners and a large niche cut into the side of the church to receive a true chapel with its altar (Ricard 1966: 165–267; Phelan 1972: 76). The mass celebrated in the true chapel allowed a large number of Indians to view the ceremony while standing in the open air. Ceremonies were then performed at each of the secondary corner chapels, thus prolonging the religious service to a proper duration by Aztec standards.

Dances in the temples had been an integral part of religion before the Conquest (Garibay 1971: vol. 1, 80–85; Martí and Prokosh-Kurath 1964: 30–83). Pedro de Gante, one of the earliest missionaries, made a conscious effort to fuse this form of worship with Catholic practices (Madsen 1967: 376–377). These dances became extremely popular and were very useful proselytizing devices. But, as another missionary, Father Acosta, wrote, they held inherent dangers: "It is not good to deprive the Indians of them but one should take care that no superstition is mixed up with them" (Ricard 1966: 183). The Church hierarchy attempted to suppress the dances as early as 1539, when the bishops prohibited them in Indian festivals honoring the patron saints of the villages (Ricard

1966: 186). Despite efforts over the past five centuries to ban the dances, these Christianized *mitotes* are still being performed, although in front of instead of within the church.

The missionaries also saw the expediency of adopting the Aztec medium of pictorial manuscripts to teach religion and prepared their own depictions of Christian doctrine. These documents (generally called Testerian writing, after Friar Jacobo de Testera, one of the earliest missionaries to New Spain) included the Ten Commandments, the Seven Sacraments, and the Articles of Faith. Pedro de Gante even composed an entire catechism in pictures (Ricard 1966: 104). The priest would stand next to the pictures drawn on canvas and use them as visual aids to illustrate the lessons he was expounding. The method turned out to be an effective pedagogical tool.

In the realm of ideology, Catholicism was also congenial to syncretism with Aztec religion. The multiplicity of Christian saints fitted very well with the Aztec pantheon. The idea that each city or town had a patron saint, who watched over it, was congruent with the Mesoamerican *calpulteotl*, a deity intimately linked to a sociopolitical entity. A number of local gods, who protected the harvests and made them plentiful, were named after the town where they were worshiped, for example, Tepoztecatl ("he of Tepoztlan"). Catholic saints were specialists who could cure specific diseases, such as St. Valentine for epilepsy and St. Fiacre for hemorrhoids (Gordon 1959: 459); sponsor professions, such as St. Barbara for artillerymen; and even guard particular crops, as St. Pantaleón protected vines against insects (Christian 1981: 43). Aztec deities were also disease specialists and profession sponsors, for example, Xipe-Totec for goldsmiths.

This congruence, together with the fact that the Spanish often built their churches on the sites of recent Aztec temples, inevitably led to syncretism. Mexican Catholicism today is a mixture of indigenous and Christian elements to such a large extent that Madsen (1967) described the folk religion as "Christo-paganism." Sahagún recognized this practice and, in the Appendix to Book Eleven of his *Historia General*, warned his fellow missionaries that heresy was taking place. His warnings were not successful: the actions he warned against thrive today. Sahagún pointed out that the Hill of Tepeacac (now Tepeyac), where the Virgin of Guadalupe was worshipped, was formerly dedicated to the Goddess Tonantzin, mother of the gods, noting the similarity of association and the Indian practice of using the name Tonantzin, "Our Mother," for the

Virgin. Sahagún called this a trick by the Devil, since the Virgin should properly have been referred to as *Dios y nantzin,* "God's and our mother" (Sahagún 1956: vol. 3, 352). The Virgin of Guadalupe is today the patron saint of Mexico and is still called Tonantzin by the Indians.

Similarly, a temple of the Aztec goddess Toci ("Our Grandmother") was replaced by the church of St. Anne, Christ's grandmother. Tianquizmanalco, the site of a feast honoring Telpochtli ("Young Man"), an avatar of Tezcatlipoca, became the site of a ceremony honoring St. John the Baptist. Sahagún (1956: vol. 3, 352, 353) pointed out that *telpochtli* also meant virgin. Since St. John also was a virgin, Sahagún recognized that the ceremony was identical to the pre-Columbian feast, but with a Christian label. In other cases, sacred rites continued to attract enormous crowds of worshippers after the Conquest under new sponsorship. The Cave of Chalco, where an apparition of Christ is worshiped, is still one of the two major pilgrimage sites in Mexico today. It was originally the locale for the worship of Oztoteotl ("the Lord of Caves"). Similarly, crosses are placed on mountain tops at the beginning of the rainy season, just where the Aztecs used to place images of the *tepictoton,* dwarf helpers of the Rain God who were believed to inhabit these mountains. Sahagún concluded:

> I believe that there are many other places in these Indies, where offerings and worship are discreetly offered to the idols, under the guise of the feasts that the Church offers to God and His Saints. This should be investigated so that the poor people should be purged of the errors under which they now labor. (Sahagún 1956: vol. 3, 354)

➡ *Sources for Understanding Aztec Medicine*

The principal written sources of information about Aztec medicine were compiled after the Conquest and were liable to be contaminated by European concepts even if written in Nahuatl. As described above, syncretism was ongoing, and material was subject to censorship to avoid charges of heresy. Our objective is to try to present an accurate picture of Aztec medicine before the Conquest. To do so it will be necessary to critically evaluate our principal sources with respect to their authenticity and reliability.

Carrasco (1982: 11) urged students of pre-Columbian cultures to practice the "hermeneutics of suspicion" before trusting documents classified as primary sources: question their nature, reliability, and intention. Because pre-Columbian writing was hieroglyphic, that is, pictorial, scribes had to be trained by the Spanish to render Nahuatl as a written language. Therefore, our primary sources, even those written by natives, have inevitably been touched and modified by Spanish culture (Carrasco 1982: 11–62). Our discussion will focus on four primary sources essential to an understanding of Aztec medicine: the works of Bernardino de Sahagún, Francisco Hernández, Hernando Ruiz de Alarcón, and Martín de la Cruz.

Sahagún

The works of Sahagún have been called "the most thorough, objective, and complete study of another culture that had ever been attempted [for its time]" (Klor de Alva 1988). They are an indispensable basis for knowledge about the Aztecs.

Fray Bernardino de Sahagún was born in 1499, but little is known about his early life. He arrived in Mexico in 1529 with a group of other Franciscan missionaries and served there until his death in 1590. For the first twenty years of his stay, he worked as a missionary, teacher, and administrator, acquiring an extensive knowledge of Nahuatl and of Aztec customs. By 1558, the religious orders were becoming disenchanted with the lack of success of their missionary efforts and feared that idolatry was invading religious practices. The Provincial of the Franciscans ordered Sahagún to "write in the Mexican language that which seemed to me useful for the indoctrination, the propagation and perpetuation of Christianity among these natives of this New Spain and as a help to the workers and the ministers who indoctrinate them" (Sahagún 1956: vol. 1, 105). There is evidence that Sahagún had begun work on this project at least ten years earlier. The project's purpose was at least three-fold: 1) to develop accurate knowledge of Aztec religion, deities, and rituals for teaching future missionaries about disguised idolatry and other pitfalls; 2) to develop a richer vocabulary for teaching Christianity to Aztecs in their own language; and 3) to develop a full record of the Aztec worldview and social organization so that, at some future time, Aztec society could be restored to its pre-Conquest splendor but under the aegis of Christianity.

In elaborating his materials Sahagún developed a unique

methodology, far ahead of its time and still good ethnography. Following Pliny's *Natural History* and other classical encyclopedic works, he posed questions in Nahuatl to a group of ten to twelve elderly informants; their answers were recorded in Nahuatl by trilingual (Nahuatl, Spanish, and Latin) student scribes. He first tried out this procedure in Tepepulco during 1558–1559 and ended up with a Nahuatl document called the *Primeros Memoriales*, which was then used to develop a more extensive questionnaire. The questions were then explored with a different group of informants in Tlatelolco during 1561. The results were expanded, amended, and revised, resulting in the Nahuatl *Madrid Códices*. In 1565, Sahagún moved to Tenochtitlan where the work was again revised and reorganized, illustrations were added, and a partial Spanish translation was prepared. This document, finished in 1569, was the basis for both the Nahuatl *Florentine Codex* and a Spanish version, the *Historia General* (López Austin 1974b; Anderson and Dibble 1982: 10–15).

Europeans were greatly interested in Aztec medicine, as shown by the enthusiastic reception of Monardes's book on Aztec medicinal plants, published in Spain in 1569 (Monardes 1925), and by the financing of Francisco Hernández's expedition to the New World by the king of Spain. Sahagún's work, which shows his constant preoccupation with the health of the body as well as that of the soul (López Austin 1974a), contains two types of medical information. Information on topics such as diseases attributed to gods, magical practices in pregnancy, and the effect of animistic forces on health arises spontaneously from answers to questions on other subjects. The other type of information, obtained when Sahagún specifically sought to record Aztec medicinal knowledge, is completely different from the rest of his work. The sections dedicated to illnesses and medicinal plants went through more changes and revisions than the rest of the *Florentine Codex*. The only informants mentioned by name are those involved in these sections; Sahagún stated that these men were "experienced doctors who cured publicly" (López Austin 1974a). López Austin (1974a) pointed out that the texts show Sahagún's influence and his censorship of as well as self-censorship by the native doctors, who knew what the friars objected to. The result is that illnesses totally related to magical or religious concepts were omitted or deleted, and the complex holistic etiology, evident in the rest of the work as well as in other sources, is simplified.

Sahagún's depth and range of coverage is not equaled in any

other source and makes it possible to correlate diverse statements in order to flesh out concepts. Thus, incidental statements in other sections of the work allow us to amplify and rectify sections where magical and religious medicinal material was deleted or omitted. Two earlier drafts of the Florentine Codex enable us to cross-check and clarify meanings by comparing parallel passages in the three versions. For example, several supernatural illnesses were listed in the *Primeros Memoriales* but deleted in subsequent drafts. Illustrations supplementing the text also facilitate our understanding. This is particularly true when clearly pre-Columbian hieroglyphs are used, as in the illustration in Book Ten of the Florentine Codex containing the hieroglyph for water to designate a fever of aquatic origin. Also, because Sahagún's is the only major primary source written in Nahuatl, it allows full use of linguistic analysis on the names of medicines and diseases and on descriptive phrases about symptoms and remedies. Translation of the names of Aztec diseases into Spanish poses the same problem in Sahagún's work as in all contemporary sources. These names may be misleading because they reflect sixteenth-century European medical knowledge, with unclear diagnoses and disease entities that differ from those of modern medicine.

A more fundamental problem is Sahagún's attitude toward Aztec religion and his censorship. The mendicant orders had become suspect when they continued to translate scripture into the vernacular in violation of an order by the Council of Trent in 1545. The Inquisition was established in Mexico in 1570 and changed the climate of opinion toward the Indians. Sahagún was suspected of heresy and of promoting paganism in his work; all his documents were confiscated and scattered (Anderson and Dibble 1982: 36). During 1578–1580 a temporary thaw in the attitude towards Sahagún's labor gave him an opportunity to prepare the final version of the Florentine Codex. He added conciliatory appendices stating that the texts concerning Aztec gods were idolatrous and false, and he further censored supernatural practices, particularly in medicine. He was not totally successful in his sanitizing efforts, as shown by numerous examples, such as retention of the entire set of illnesses relating to animistic forces. Sahagún, like all his contemporaries, also believed in magic and was unable to eliminate it completely.

Thus, Sahagún's sins were mostly sins of omission, the deletions of Aztec concepts, rather than sins of commission, the insertion of European concepts. His insertions of European concepts are usually clearly labeled in appendices and commentaries. What the Sahaguntine corpus does is give an impression of less supernatural involvement in Aztec medicine than there was in reality. We will use Sahagún as the first and most valid source, and as a touchstone for the validity of data from other sources, whose reliability and credibility are greatly enhanced if they are in accord with Sahagún.

De la Cruz (Badianus Codex)

Another product of the conflict over the nature of the Indians and of Indian policy was the herbal written by Martín de la Cruz in 1552.[3] By 1552, the Colegio de la Santa Cruz in Tlatelolco was in trouble. It had aroused the ill will of those Spaniards who felt that the Indians were inferior and should not be educated. Two of its main political supporters were gone; Viceroy Antonio de Mendoza had left for Peru, and Archbishop Juan de Zumárraga had died in 1548. The school had lost a royal financial subsidy because King Charles V had left Spain in charge of sixteen-year-old Prince Philip II, who had neglected to renew it (Somolinos d'Ardois 1964: 301–302). As a political move, Francisco de Mendoza, the Viceroy's son, solicited an herbal as a gift to be presented to Charles V in order to "recommend the college to the king," as the dedication states (de la Cruz 1964: 199). The herbal was to be a tool in lobbying for the view that Indians were "human," capable of being educated, and possessors of a worthwhile culture.

The herbal was prepared in haste and written in Nahuatl by an Aztec doctor, Martín de la Cruz, about whom nothing else is known. It was subsequently translated into Latin by Juan Badiano, an Aztec professor at Tlatelolco. The work was illustrated in color and probably modeled after such medical herbals as Johan von Kaub's *Hortus Sanitatis* (Somolinos d'Ardois 1964: 312, 321). It was sent to Europe, and Somolinos d'Ardois (1964: 303–304) felt that it was seen in Spain by Prince Philip but not by King Charles. It is not clear if it influenced European opinion about the merits of Aztec civilization, since it was not cited in any work published during the next three centuries. De la Cruz introduced a number of elements into his herbal to meet and demonstrate European cultural standards. It was translated into Latin because that was the language used by all "civilized" people. Citations of Pliny and the prescription of contemporaneous European medical practices, such as the use of bezoars (inorganic solids usually found in the alimentary tract of animals and used in Europe as an antidote to poisons), gem stones, complex mixtures of substances, and coprotherapy, were also designed to demonstrate knowledge of European culture.

The value of the *Badianus* as a primary source for knowledge of Aztec medicine is diminished precisely because of these efforts to show European sophistication. What is authentic Aztec is difficult to distinguish from what is European or distorted to fit European predilections. For example, European contamination is clear in the prescription of bezoars from nineteen different animals (del Pozo 1964: 340) and in several medical applications of roosters, which are not native to the New World (de la Cruz 1964: 189, 199, 209, 217, 225). Also, del Pozo (1964: 334) pointed out that descriptions by de la Cruz of the pathological effects of bile and the notion of black bile leading to melancholia owe more to Galen and Hippocrates than to native concepts. The *Badianus* cites Pliny (de la Cruz, 1964: 169) on the use of *alectoria,* a stone found in the crop of roosters, to alleviate thirst. That Pliny was available to de la Cruz is indicated by the listing of Pliny's *Natural History* in a 1572 inventory of the school's library (Anonymous 1886–1892: vol. 5, 260). Other items parallel but do not cite Pliny directly. For example, the herbal's prescribed treatment for comitial disease (epilepsy) included a mixture containing deer's horn (de la Cruz 1964: 209); Pliny (1949–1962: vol. 3, 83—book 8, chap. 50) had stated that the smell of burned deer's horn arrests the attacks of epilepsy. The *Badianus* says that burnt deerskin helps a woman give birth (de la Cruz 1964: 217); Viesca Treviño and de la Peña (1974) pointed out that this is a magical prescription using the principle of "like produces like," since does supposedly give birth very easily, an idea found in Pliny (1949–1962: vol. 3, 81—book 8, ch. 50). Pliny had written that does ate hartwort to ease giving birth. The *Badianus* also includes a multitude of magical remedies, in strong contrast to the medical sections in Sahagún's works. Pliny, who was not a doctor, recommended magic, coprotherapy, and complex mixtures as remedies to an extent excessive even by the standards of his time, as can be seen by comparison with the procedures recommended by Celsus (1935–1938), Pliny's contemporary.

Since both Europeans and Aztecs believed in magic, we will have to distinguish between native and foreign magical remedies whenever possible on the basis of congruence with their different worldviews. Del Pozo (1964: 333) noted the absence of de la Cruz from a list of Sahagún's informants and argued that Sahagún may have excluded him because he felt that de la Cruz was too influenced by European medicine to be representative of Aztec medicine, or mixed his medical data excessively with magical practices.

The lack of Nahuatl names for diseases and remedies in the Latin *Badianus* makes it difficult to correlate and compare it with other sources. The problem of identifying Aztec diseases is exacerbated because Badiano was forced to search for ailments recognized by Pliny in first-century Rome that were analogous to those listed by de la Cruz. According to del Pozo (1964: 333), the use of Latin names of diseases did not imply a knowledge of European medicine, since sixteenth-century diagnosis was erratic. He cautioned that we should be careful in attributing to those illness designations the precision which they enjoy today after centuries of medical advances. Del Pozo (1964: 330) stated that even a layperson can diagnose epilepsy, but that other diseases cited in the *Badianus*, such as *podagra* ("gout"), *mentagra*, a skin disease, and *condyloma*, a wartlike growth, are much more ambiguous. On occasion, Badiano had to resort to strange analogies in order to match some diseases, such as his use of *abderitatis mentis* ("Abdera-like mind") to designate a form of mental stupor,[4] or *siriasis* to denote a child's burn (de la Cruz 1964: 221), when the classical definition is an inflammation of the meninges causing a high fever (Temkin 1956: 124). The use of Latin, apart from distorting Aztec illnesses to fit a Roman mold, denies us the full benefit of a linguistic approach to elucidate the etymology of disease names, symptoms, and remedies. For example, *calor* ("fever") is much less informative than *atonahuiztli* ("aquatic fever") or *iztac totonqui* ("white fever"). Finally, the use of Latin obscures the presence of foreign elements. When Badiano wrote *frumento* ("wheat"), was he referring to the European product or to corn, for which there was no Latin word?

For our purposes, the *Badianus Codex* is best used with caution and in conjunction with other sources such as Sahagún. The realistic illustrations make botanical identification of many plant species possible, but the pictures and text lack supplementary Aztec hieroglyphs, which would have been quite illuminating. The herbal format precludes the depth and context found in Sahagún. Because we have only the Latin translation, we can only guess at the Nahuatl equivalent for the names of illnesses. A number of illnesses, such as the effects of lightning strikes and fear, can be identified by context and by cross-references to other sources. The high proportion of magical remedies is a useful counterweight to the view presented by Sahagún's informants. Because the herbal's principal purpose was to impress European royalty, we cannot regard it as an accurate and encyclopedic view of Aztec medicine.

Ruiz de Alarcón

Hernando Ruiz de Alarcón (1587?–1646) was the second son of a distinguished Spanish family living in Taxco. Because he was born in Mexico of pure Spanish parentage, he was a *criollo* (creole). His elder brother was the eminent playwright Juan Ruiz de Alarcón. Hernando graduated from the University of Mexico in 1606, took holy orders, and became a parish priest in Guerrero (Ruiz de Alarcón 1982: 11–14).

Ruiz de Alarcón grew up at a time when the bright promise of the Conquest was fading. Diseases and terrible working conditions resulted in a precipitous decline in the native population, and consequently in the economy. It also had become clear that conversion of the Indians would be more difficult than expected. King Philip II had established the Holy Inquisition in Mexico in 1570. Although jurisdiction over the Indians was officially reserved to the bishops, the Inquisition investigated instances of native idolatry and superstition. It looked with special suspicion on the mixing of Christian and Aztec patterns of belief and the Indians' use of botanical hallucinogens for divination, which smacked of witchcraft (Ruiz de Alarcón 1984: 6–7; Aguirre Beltrán 1963: 115–162).

In 1614, the Inquisition learned that Ruiz de Alarcón was conducting his own auto-da-fés and punishing Indians for idolatry. The Inquisition admired his zeal and co-opted him. He was appointed as an ecclesiastical judge in 1617 and began to inform the Inquisition about instances of idolatry and superstition among Indians in his region. In his official capacity he served as both prosecutor and judge. He often obtained confessions by using informers, relatives, and the threat of long prison sentences (Ruiz de Alarcón 1982: 20; 1984: 7). His *Treatise on Superstitions*, in 1629, was the result of these endeavors.

Alarcón's purpose in writing the book was to help cast out paganism and syncretism in native practices by recording beliefs and activities he believed to be pagan. He acknowledged the danger that, by recording these practices, he might encourage them, but argued that it was much more urgent to help other clerics recognize and extirpate these activities (Ruiz de Alarcón 1984: 7). In this respect Alarcón's book resembles Sahagún's work, but it differs markedly in its attitude toward the natives. Far from hoping to rekindle the Aztec state on a Christian foundation, Ruiz de Alarcón, like most Spanish creoles, was contemptuous of all Indians. He considered native doctors charlatans who fooled other Indians with spells and tricks. When sorcery actually worked, as he believed it sometimes did, then the sorcerer must be in league with the Devil (Ruiz de Alarcón 1982: 22). He adopted an adversarial tone in both collecting and presenting the evidence. Rather than a sympathetic and comprehensive ethnography like Sahagún's work, he prepared a lawyer's brief.

The *Treatise* was not designed as a comprehensive survey of Aztec beliefs or even as a survey of medical practices. Only a few medicinal plants are mentioned—fewer than thirty—which contrasts radically with our other sources. The cures he described involved hallucinogens, mostly symbolic objects such as water or *copal* (incense), and magical incantations rather than real materia medica. He leaves the impression that Aztec medicine was *primarily* magical and rarely attributed illness to naturalistic or religious causes:

> The ones suffering from protracted illnesses and from those which, being confirmed, are considered by the doctors to be incurable, such as are the consumptives and the tuberculars, etc., seeing that they do not improve with ordinary medicines, then attribute their illness and sickness to witchcraft and at the same time consider it certain that they will never become well if the one who bewitched them does not cure them or does not want them to get well. (Ruiz de Alarcón 1984: 63)

Ruiz de Alarcón's focus on magical treatments automatically excluded empirical remedies for natural illnesses or for those that were not chronic. This may be the reason that diseases caused by "heat" or "cold" are not mentioned in this source but are found in the others.

Ruiz de Alarcón's *Treatise* extends and supplements concepts of the illnesses ascribed to animistic forces found in other sources as well as the magical and divinatory aspects of Aztec medicine. It is a unique source for magical incantations used in medicine. Although these incantations were surely used often, other sources have not recorded them. The incantations are problematic because they are written in an abstruse style full of obscure metaphors and difficult to translate and understand.

A parallel source is Jacinto de la Serna's *Manual de Ministros* (de la Serna 1953). Jacinto de la Serna was born in 1600 and died in 1681. In his lifetime he became a Doctor of Theology and three times served as Rector of the University of Mexico. He wrote the *Manual de Ministros*, based upon the first draft of Ruiz de Alarcón's manuscript, for the use of the clergy in eradicating idolatry and superstition among the Indians. It was completed in 1656. Approximately seventy-five percent of de la Serna's book is based on Ruiz de Alarcón; some forty percent is direct quotations (Ruiz de Alarcón 1982: 53). The remainder is based on de la Serna's own personal observations.

Both the *Treatise* and the *Manual* are too limited in scope and purpose to be accurate guides to Aztec medicine. They are most useful as adjuncts that can extend and supplement more comprehensive and accurate sources. They, too, augment the information on magical cures minimized in Sahagún.

Hernández

Francisco Hernández's *Historia Natural de la Nueva España* differs from the previous sources in both the training of the author and the book's genesis. It did not arise as part of the power struggle between the secular and regular orders for control of the Indians.

The discovery of the New World had aroused an enormous amount of interest in Europe. Veridical reports of expeditions were mixed with wild stories of monsters and mythical animals. Of particular interest were reports detailing an abundant number of new medicines and unusual foods. Concern about the epidemic of syphilis, believed to have originated in the New World, led to a massive importation of two New World remedies, holy wood (*Guaiacum sanctum*) and sarsaparilla (*Smilax* sp.), on the theory that each land provides the cures for its own special diseases. King Philip II appointed Francisco Hernández, one of his royal physicians, *Protomédico*[5] of the Indies and sent him on a scientific mission to the New World. Hernández was directed to make a comprehensive study of the medicinal plants of New Spain and Peru, including their use, how and where they grew, and an estimate of their effectiveness. Execution of this mission culminated in the *Historia*. The magnitude of the task was so great that, even though the viceroys were ordered to assist Hernández in his work, he never managed to get to Peru (Somolinos d'Ardois 1960: 144–152).

Francisco Hernández was probably born in 1517 near Toledo

and was trained in the classical Hippocratic-Galenic tradition at the medical school in Alcalá de Henares (Somolinos d'Ardois 1960: 101–103). After graduation, he practiced medicine in various cities and achieved the status of *médico de cámara*, royal physician. Although this appointment did not necessarily imply that he personally attended the king, it was the highest level of recognition and implied certain official duties. The importance given by the king to the scientific mission is demonstrated by his choice; at the time of his appointment Hernández was one of the top physicians in Spain and at the peak of his career.

Hernández left Spain in September 1570, arrived in Mexico in February 1572, and remained there until 1577. He went directly to old Indian doctors and questioned them through an interpreter. None of his informants are cited by name. Native artists were employed to prepare illustrations of plants to be discussed in the *Historia*. Hernández often tried to verify received information by testing the plants on himself or on patients in hospitals to which he had access. He also conducted several expeditions to various areas of Mexico to collect local flora and interrogate informants outside Mexico City. One of his most important trips was to Huaxtepec, the site of a royal botanical garden established by the emperor Motecuhzoma I in 1453, where doctors had been passing down information from father to son for generations (Somolinos d'Ardois 1960: 196–202).

In 1576, Hernández's book, the most extensive compilation of medicinal plants in all our sources, was sent to Philip II. The king reportedly greatly admired the *Historia Natural* for the beauty of the illustrations and the comprehensiveness of the data. Hernández returned to Spain in 1577. Unfortunately, his work was not published before his death in 1587, and important portions of the manuscript were lost.

Philip appointed a Neapolitan physician, Nardo Antonio Recci, to edit Hernández's work for publication. According to Somolinos d'Ardois (1960:282), Recci knew nothing about America, did not understand the value of Hernández's work, and mutilated it, deleting comments he considered superfluous, such as all personal references and all that did not pertain to medicine. Several copies of the edited manuscript were made. One ended up back in Mexico at Huaxtepec, and one was taken by Recci to Naples and became the basis, in 1648, for the first publication of Hernández's work. Although Recci diminished the value of the book, he was also its savior, because Hernández's original manuscript, including the irreplaceable native illustrations, was destroyed in the fire at the royal library of the Escorial in 1671 (Somolinos d'Ardois 1960: 282).

Hernández was so imbued with Galenic theory and training that it colored his data gathering and understanding of Aztec medicine. Somolinos d'Ardois (1960: 175) felt that Hernández impartially both criticized the deficiencies of the natives and praised their empirical knowledge while disagreeing with practically all of Aztec medicinal theory. Hernández's attitude toward the Indians had neither the hostility of most other contemporaneous observers nor contempt like that displayed by Ruiz de Alarcón. He was full of praise for the extent of the Indians' botanical and taxonomical knowledge:

> I marveled, in this and in innumerable other herbs, which are nameless among us, how in the Indies, where people are so uncultured and barbaric, there are so many herbs, some with known uses and some without, but there is almost none, which is not known to them and given a particular name. (Hernández 1959–1985: vol. 5, 425)

Hernández classified all plants according to his Galenic training as hot or cold and wet or dry and, again in Galenic tradition, often quantified these qualities by degree of hotness or wetness. This intrusion makes it difficult to find out how the Aztecs themselves classified particular plants except, as is often the case, when Hernández gives the native classification and then disagrees with it. For example, in referring to *cececpatli* ("cold medicine"), classified by Hernández as hot to the third degree, he stated:

> The Indians of *Atotonilco* say that it cures fevers, when eaten or applied locally but since this appears contrary to the nature of this herb, I feel that it is more probable that other effects, more in accordance with its nature, are derived from its temperament. (Hernández 1959: vol. 2, 235)

Hernández's training in the Galenic theory of humors and of curing by correcting humoral imbalances prevented him from understanding Aztec etiologic concepts different from his own. For example:

> [*achilto*] They say that the sap put into the nose cures headaches, which seems senseless since its nature is cold and

glutinous unless the pain is due to a hot cause. (Hernández 1959: vol. 1, 132)

However, the Aztecs believed that headaches were caused by excess blood in the head and cured by its removal, rather than due to an imbalance of hot and cold, a Galenic concept that Hernández could not abandon.

In discussing the uses of the plant *chichioalpatli*, Hernández (1959: vol. 1, 188) commented that he knew many other uses, omitted for the sake of brevity, unknown to the Indians because they knew only a few for each plant. Many other plant descriptions are accompanied by a laundry list of uses. It is not clear whether they truly reflect native practice or include Hernández's supplements based on his own precepts. Other examples evidence European penetration, particularly in the treatment of conditions that were not native, such as poisoning by mercury vapor (Hernández 1959: vol. 1, 384) and siriasis, an infantile inflammation and fever (Hernández 1959: vol. 1, 228), or in the use of European remedies such as bezoars (Hernández 1959: vol. 2, 307).

The advantage of Hernández's work, despite alterations, is its comprehensive scope, especially in the number of listed plants, 3,076 according to del Pozo (1964: 337); Sahagún and de la Cruz described a much smaller number. As a trained observer, Hernández provided very useful extended commentary, particularly for food plants. However, the *Historia Natural*, like the *Badianus*, has the disadvantage of having been written in Latin, depriving us of Nahuatl etymology. Moreover, its translation into Spanish added distortions of meaning that accompany all translations. Consequently, we do not get the Nahuatl names for illnesses except indirectly when a listed plant name has the suffix *patli* ("medicine for"), and we lose the opportunity to apply linguistic analysis to the context, symptoms, and remedies for these illnesses. Also, the *Historia* deals only with natural products, because of Recci's elimination of extraneous matters. Since Hernández was unable to understand native etiologies, we are also deprived of the kind of peripheral and incidental information about health and disease that we get in Sahagún's extended ethnography.

Hernández's usefulness as a source for evaluating the empirical validity of Aztec medicines is diminished because most of the plants he listed have not been identified botanically, an essential first step for a modern investigation. The first publication (Hernández 1942–1946) contained identifications, but was only a partial translation of the original work. The complete translation

(Hernández 1959) did not include plant identifications, which were reserved for a later volume (Hernández 1959–1985: vol. 7). Unfortunately, those identifications were culled from older publications and do not represent modern standards. Also, a potential pitfall is the fact that Hernández included plants native to Peru and the Philippines but alien to Mexico.

The *Historia Natural*, like the *Badianus*, is most trustworthy when it confirms or extends concepts found in other sources, principally Sahagún. With proper precautions to try to exclude European influence and bias, the numerous examples in the *Historia* can provide information about therapeutics and can be used to test hypotheses about etiology. The overall effect of the *Historia Natural* is to show Aztec medicine, somewhat paradoxically, as both more naturalistic and to some extent more magical than it probably was. Most noticeably absent are notions about gods and the supernatural as causes of disease.

The first publication of Hernández's work was due to a priest, Francisco Ximénez, who worked at Huaxtepec and who had one of Recci's edited copies of the book. Ximénez published an extract of the *Historia* and added some commentary (Ximénez 1888). His version has the same advantages and disadvantages as the fuller account.

No single source represents fairly and accurately what Aztec medicine was like. We will show throughout the book that Aztec medicine involved naturalistic, magical, and religious aspects. We can get a much better and more accurate picture of Aztec medicine by judiciously combining all four of our sources. Sahagún's etiology combines naturalism and religion but little magic. The *Badianus* and Ruiz de Alarcón contain much magic and some religious etiology but few examples of natural causes. Hernández presents primarily a naturalistic image with some magical aspects. Taken together these sources enable us to present a credible view of pre-Columbian Aztec medicine.

Methodology

Assessing and Discounting European Influences

Our main goal is to sketch an accurate picture of Aztec medicine at the time of the first European contact. This includes the related and complementary aims of disentangling and clarifying the extent of European contamination in the sources and determining the authenticity and completeness of the data. As we have seen, exclusive reliance on any one source will produce an idiosyncratic image of the nature of Aztec medicine, ranging from predominantly naturalistic at one extreme to predominantly supernatural at the other.

Views differ on the extent of European contamination of Aztec sources. On the one hand, Kidwell (1983) found very little evidence of transmission of knowledge (other than about medicinal herbs) from Aztec to European sources, or from European to Aztec sources, during the first century of contact. On the other hand, Kay (1979) and Foster (1978a, b; 1986; 1987), among others, argued that the extent of contamination by Spanish medical theory is so great, even in Sahagún, that in some areas such as classification of diseases one cannot discern what is genuinely native. Foster claimed that the Hispanicized Indian scholars who wrote the available documents had been trained using European books and, in particular, had so absorbed Hippocratic-Galenic humoral theory that references in native sources to the "hot" or "cold" nature of a remedy should not be considered aboriginal. Foster also referred to the presence of Spanish words, such as *xerencatica* (*jeringa*—"syringe"), *trigo* ("wheat"), and *hora* ("hour"), in the original Nahuatl text of Sahagún's informants as further evidence of contamination.

Certainly the mere statement that a plant is hot or cold does not conclusively indicate either native or European origin. However, the sources contain some indications of the existence of aboriginal "hot" and "cold" physiological concepts, as pointed out by López Austin (1988: vol. 1, 279):

> Weightier evidence has a more profound character: the etymological analysis of words that reveals the existence of a system; concepts of dichotomy that extend far beyond the narrow field of health and sickness; the logical relatedness of elements within a taxonomic system and a world view, etc.

The presence or absence of borrowed words does not necessarily prove deep cultural influence. It took over a century of intense contact with Spanish before the natives began to use Spanish nouns extensively, and these primarily represented concepts for which terms were not available in Nahuatl, for example, Christian religious terminology such as *angelome* ("angel") and names plants, animals, and foods introduced by the Spanish, such as *cauallo* (*caballo* = "horse") and *castillan tlaolli* ("castillian corn" = "wheat") (Karttunen 1985: 11, 54). Such a circumstance would explain the instances cited by Foster. Very few words for abstractions were borrowed in the sixteenth century, and nouns were borrowed well before verbs or other parts of speech (Karttunen 1985: 53, 56). Therefore, the pure Aztec origin of a physiological or medical concept is more likely if it is expressed in Nahuatl with other parts of speech in addition to nouns.

We will follow López Austin's lead in linguistic analysis of words and expressions as a way to understand Aztec ideas and beliefs and to identify authentic Aztec concepts. The characteristics of Nahuatl itself assist this endeavor:

> Nahuatl in the period just before the Conquest was a language in which the composition of words was highly descriptive: there was an astonishing connection between the actual meaning of the words and the elements that composed them. The possibility of forming new words under the strict principles of composition allowed simultaneously for expressive and descriptive richness, conceptual transparency, and a sensibility to historical change; and not having the rigidity of a lexicon difficult to modify, the language could respond to changing circumstances with words coined in the act. This easy response to innovation tended to make it reject terms from foreign languages. . . . I believe that few languages can give a greater guarantee of approximating history where philological analysis has been used as an aid to the study of the ideology of a people. (López Austin 1988: vol. 1, 21)

Apart from linguistic analysis, other types of evidence support the authenticity and reliability of information found in sources. Instances in which a Spaniard clearly states that a particular Aztec concept is wrong clearly distinguish what is native and what is European. As such they can be more informative and illuminating than straightforward descriptions of Aztec practices. For example, Hernández listed humoral qualities for most of the remedies in his work but clearly deduced them from Galenic principles; plants with a strong odor or taste or a bitter flavor he classified as "hot," while plants with little odor or taste or a sweet flavor, he classified as "cold." Significantly, when he criticized native doctors for misclassifying or misusing certain plants, his criticisms make no sense unless the Aztecs themselves applied "hot" or "cold" labels but according to classification principles different from his own; there is no reason that two systems developed independently should give identical results. The following, taken from sixty-one examples of Hernández's classifications and comments, illustrate these points:

a) *memeyas* (vol. 1, 323) "They are almost all of a hot and dry nature, even though after the milky juice is squeezed out they don't show a trace of heat, but the Mexicans still claim that they are cold and useful against fevers."

b) *hoacxochitl* (vol. 1, 383) "The *Itztolucanenses* . . . say that its bark is hot and combats cold and tumors but since I found it insipid and with no trace of heat, I classify it as cold."

c) *tlachinolxihuitl* (vol. 2, 52) ". . . hot, dry and fragrant. . . . They say that it is used against erysipelas, but this is unreasonable because it is hot."

d) As shown in chapter 8, *coacihuiztli* was caused by Tlaloc and was a "cold" disease for which a "hot" remedy would be indicated. Hernández (vol. 1, 223, 224, 340, 358, 359, 361) lists six *coacihuizpatli* ("medicine for coacihuiztli") and classifies three as hot and three as cold. Clearly, his classification scheme differs from that of the natives, for whom all six are hot, and he points out his disagreement (vol. 1, 223): "The leaves have no taste [therefore classified as cold] but nevertheless the Indians say that applied locally they cure pains." (Hernández 1959)

Similarly, Ruiz de Alarcón (1984: 203–204; de la Serna 1953: 285) expressed his amazement that, contrary to the rest of humanity, Aztecs rested before working instead of after. Although this did not make sense in European terms, it is consistent with the existence of the animistic force, *tonalli*, which heats up during work and has to be cooled beforehand (López Austin 1988: vol. 1, 260–262).

Credibility is enhanced when the same concept is expressed in several different sources, demonstrating widespread knowledge of the topic and shared assumptions. This is particularly important when sources with different biases concur. Among the various areas of knowledge, the Aztec worldview and religion were less subject to European contamination than any other. Therefore, if the statement of a particular fact or belief is compatible with the native myths and worldview, it gains greater credibility and validity. As we will see, the belief in animistic forces, particularly tonalli, will be particularly useful. Tonalli was a "soul" located in a person's head that provided a vital force. Clearly, tonalli was authentically Aztec; it fits with shamanic elements in Aztec religion, such as a belief in the existence of multiple souls. This belief was widespread in the New World but alien to the Spanish experience. Tonalli is enormously useful and has wide-ranging application, for example, to reconcile and explain items as diverse as various diseases, astrological fates, the capture of warriors, the legitimacy of rulership, the need for moderation in sex, and the failure of dyes to work in the presence of twins. The abundant evidence about tonalli in diverse sources clearly marks it as an exclusively native concept. If a statement in a particular source can be explained by or related to tonalli, its authenticity is strengthened. For example, as illustrated later, even though European humoral medicine encompassed the idea of "hot-cold," many "hot-cold" designations in Aztec sources can be related to tonalli and therefore probably refer to a native pre-Columbian classification rather than to Galen.

Peasant societies are very conservative. Cultural and linguistic traits persist with little change for long periods of time. Modern ethnographic data can therefore be used to amplify pre-Columbian notions, and knowledge about the Aztecs can be used to explain current practices. For example, some viral warts are called *mezquinos* ("misers") in Mexico today. Foster (1985; 1987) cited this name of Arabic origin as evidence for the influence of Spanish medicine on Mexican folk medicine. This name for warts was also found by Kay (1977b: 142) in Arizona, where medicinal concepts are also claimed to be of purely Spanish origin. However, an alternate explanation is possible. López Austin (1988: vol. 1, 23) pointed out that these warts were also called *tzotzocatl* ("miser") in sixteenth-century Nahuatl, antedating the borrowed Spanish noun, but that no etiology was cited. Madsen's (1960: 167) study of Tecospa, an Aztec village near Mexico City, provided an explanation. According to Tecospans today, warts are of a cold nature and are caused by selfish feelings. People who have them want to get rid of them so that their selfishness will not be revealed. Thus, the designation may have an Aztec rather than a Spanish origin.

Empirical Evaluation

Most studies of Aztec medicine have focused on the religious and magical characteristics of treatments. However, the successful use of ethnographic and ethnohistoric sources for the identification of pharmacologically active hallucinogens (Schultes and Hofmann 1980), as well as recent advances in ethnobotany (Berlin,

Breedlove, and Raven 1974), have demonstrated that native people possess detailed and accurate knowledge of the natural world. Since the Aztecs were accurate observers of nature, it is of interest to evaluate Aztec medicine objectively by the techniques of modern science. A distinction must be made between emic and etic aspects. Emic aspects of a culture are the native's own perceptions, concepts, thoughts, and values. Etic aspects are native behaviors and concepts that can be objectively verified by an outside observer. Most evaluations of folk medicine in the past have taken the etic approach, judging folk medicines effective or ineffective solely by Western biomedical standards. A fair evaluation and richer view of folk medicine should include an empirical assessment of emic aspects; that is, did remedies work on strictly Aztec terms?

Our approach to the empirical validation of the emic aspects of Aztec medicine requires the identification of testable hypotheses derived from Aztec etiologic ideas. In some cases, etiologic ideas are clearly stated; in others, they must be uncovered by preliminary detective work. Given a particular etiology, the required cure can be logically deduced. To evaluate a fever remedy in emic terms, we must determine whether it could have produced the particular effects expected by the Aztecs. And to decide objectively whether a prescribed plant remedy can act as, say, a diuretic, we must first be certain of its botanical identification. With the plant's modern scientific name we can search the chemical, biological, and medical literature to identify its chemical components, note the physiological and pharmacological activities of these components, and compare them to the effects an Aztec physician would logically derive from his own etiologic premises.

Emic evaluation of Aztec medicine using this methodology shows that the Aztecs correctly observed the physiological effects of plants and could produce the desired effects on the basis of the underlying Aztec theories of physiology and etiology.

Botanical identification of Aztec medicinal plants can also be used in other ways. The morphology and chemical constituents that give a plant its characteristic odor and taste often provide a rationale for some of the magical uses of the plant. Patterns of characteristics exhibited by plants prescribed for particular ailments can be used to derive etiologic hypotheses, which can then be tested independently. In this respect, the incomplete identification of species listed in Hernández's *Historia Natural* is especially regrettable and severely restricts its value as a source.

Recent surveys have shown that Western medicines still contain a considerable percentage of natural organic products—estimates range from thirty-five to fifty percent. Many of these natural substances have come from the scientific study of traditional remedies (Farnsworth and Bingel 1977). Since there are several hundred thousand different species of the higher plants, a random search for beneficial medicinal products would be very expensive and yield meager results. Etic evaluation of ethnomedicine can improve the success rate considerably. Since an evaluation of the emic component of Aztec medicine has shown the Aztecs to be very accurate observers of the natural world, one would expect them to have discovered a number of remedies valid by etic Western standards.

Indeed, this methodology found Aztec medicine to be quite effective by etic standards, as shown by the evaluation of Aztec medicinal herbs in Appendix B. Over half produce the claimed effects according to both folk reports and experimental evidence. The Aztecs clearly were keen observers of nature and used empirically derived remedies.

In this work, we have tried to be critical of sources, using intercomparisons to sort out error and to derive a reliable description of Aztec medicine, diet, and health. We use the work of Sahagún whenever possible as a touchstone for the authenticity of information in other sources. However, such authentication is not possible for some material, such as magical incantations, found only in other sources. We use such material as valuable evidence of the extent of magical beliefs in the Aztec culture.

Chapter

2

Aztec
Religion,
Worldview,
and Medicine

The ethnographic work of Bernardino de Sahagún and some early
writings of Spanish missionaries, which described Aztec religion
for the purpose of preventing syncretism and heresy, are rich
sources for the study of the Aztec religion and worldview. Although we speak of "Aztec" religion, much of what follows also
applies to other parts of Mesoamerica because of a regional commonality of beliefs, with some local variations (Orellana 1987: 23–
42; Ortiz de Montellano 1989b). Aztec religion was a state-
ecclesiastical institution, that is, with priests, temples, and a formal hierarchy, but differed from most other ecclesiastical religions
in the world in its retention of shamanic elements (Furst 1976: 4,
57–86). These shamanic elements will be discussed further below
and in chapter 7.

The Aztecs held the unique view that the structure and function of the human body paralleled the structure and organization
of the universe. The body was also linked to the universe; astro-
nomical events could affect bodily functions and, conversely, human behavior could affect the equilibrium and stability of the
universe. One of the key duties imposed on man was to maintain
the existence of the universe through the performance of rituals,
for example, supplying energy to the sun by human sacrifice.

Aztec origin myths fostered the belief that man "owed" the

gods because divine sacrifices had created the world and man himself. This ideology led to a system of ethics and a concept of good and evil based on obedience and the performance of duty (Read 1986). The Aztecs divided the universe into opposite but complementary halves, a duality that extended to the human body: health and happiness were the result of dietary balance and moderate behavior. Thus, the ethic of performance of duty and moderation was reinforced by the physical consequences of its violation. The body was hostage to adherence to behavioral norms. Deviation from duty or excessive actions (particularly of a sexual nature) were considered major causes of illness.

In this chapter we discuss Aztec religious beliefs, concepts, practices, and structures. Emphasis is placed on those that had a special bearing on human physiology, health, and medicine: the duality of the cosmos; creation, destruction, and human sacrifice; man as microcosm; complexes of deities; the preservation of cosmic order as a justification for social controls; multiple souls; social and religious stratification; and shamanism.

▬ Duality of the Cosmos and Its Constituents

The concept of cosmic duality was one of the key organizing factors in Aztec culture. According to the Aztecs, the cosmos was divided into two great parts, which could explain its functioning, organization, and movement. The parts were paired, simultaneously opposed and complementary in a manner analogous to the Eastern concept of yin and yang. A horizontal plane separated the heavens (the Great Father) from the earth-underworld (the Great Mother). This basic division carried with it a number of pairs of opposites:

Mother	*Father*
female	male
cold	hot
down	up
underworld	heavens
wet	dry
dark	light
night	day
water	fire
death	life

The first humans, according to myth, had been created from both celestial and earthly matter, a mixture of opposing forces that had to be maintained in balance. This justified an ethic of moderation and led to a belief in disequilibrium as a cause of illness. As we shall see, this belief facilitated the syncretic assimilation of the Hippocratic idea of an essential need for a balance of humors (hot and cold) introduced by the Spanish conquerors.

The idea of duality extended to other areas. Mythologically, the prime creator gods were a male and female pair—Ometecuhtli ("The Lord of Duality") and Omecihuatl ("The Lady of Duality")— who resided on Omeyocan ("the Place of Duality"). The male-female dichotomy was found in various other deities, among them the gods of water, corn, and death. Duality also extended to the organization of the Aztec state: power was divided between two top rulers, the *tlatoani* ("chief speaker"), the military ruler who represented the gods' male attributes, and the *cihuacoatl* ("snake woman"), the religious leader and administrator whose name tied him to the feminine aspects of divinity (López Austin 1988: vol. 1, 76).

Aztec deities also expressed opposite characteristics within themselves. For example, a god could both cause disease and cure it, nurture crops and destroy them, or provoke sexual sin and forgive it. The concept of creation and destruction as opposite facets of the same entity can best be seen in the goddess Coatlicue ("Serpent Skirt") and the earth-mother complex to which she belonged. Coatlicue was the mother of the sun, the moon, and the stars. According to myth, she magically conceived Huitzilopochtli, the sun god, at a hill called Coatepec ("hill of the serpent"). Her daughter, Coyolxauhqui, the moon goddess, plotted with her brothers, the Centzon Huitznahua (who represent the stars), to kill the sun as soon as he was born. Huitzilopochtli was forewarned and was thus born fully grown and armed. He defeated his siblings, just as the rising sun "defeats" the stars and the moon. He dismembered Coyolxauhqui and threw her and her brothers down to the bottom of Coatepec (Sahagún 1950–1969: book 3, 1–2).

All goddesses of the earth-mother complex shared to some degree an involvement with maternal fertility and death. The earth was simultaneously the womb and tomb of all life. The Aztecs thought of the earth as a great monster that swallowed the sun each sunset, threatening all life, but also provided the food that sustained life. The fertility-death combination can also be seen in the iconography of the great statue of Coatlicue discovered in 1790 (fig. 2.1):

Figure 2.1.   Statue of Coatlicue, Goddess of Earth and Fertility.

In accordance with her name, Coatlicue wears a skirt fashioned of braided serpents which is secured by another serpent in the form of a belt. A necklace of alternating human hands and hearts with a human skull or death's head as a pendant partially covers the goddess' breasts. Her feet and hands are armed with claws, since she is the insatiable deity who feeds on the corpses of men. That is why she was also called "the devourer of filth." Her breasts hang flaccid, for she has nursed both the gods and mankind, since they are all her children. Hence she was known at [sic] Tonantzin, "our mother," Teteoinan, "the mother of the gods," and Toci, "our grandmother." Her head has been severed from her body, and from the neck flow two streams of blood which take the form of two serpents portrayed in profile, their touching heads forming a fantastic face. Down the back of the goddess hangs an ornament of strips of red leather tipped with small shells, the decoration characteristic of earth gods. (Caso 1958: 53–54)

= Creation, Destruction, and Human Sacrifice

The concept of intertwined creation and destruction is consistent with the fundamental cosmological beliefs of Mesoamerica (Read 1987). The Aztecs believed that the world had been created and destroyed several times. Each of these ages ("suns") possessed a different set of characteristics and was ruled by a particular deity. At some stage in each "sun," a disequilibrium led to chaos, then a characteristically cataclysmic destruction, and finally a new creation with a new patron deity (López Austin 1988: vol. 1, 67) that again displayed divine duality. During the first era, Ocelotonatiuh, the earth was ruled by Tezcatlipoca and inhabited by giants who ate acorns. It was destroyed by jaguars on a day 4-ocelot in the Aztec ritual calendar. During the second era, Ehecatonatiuh, the earth was ruled by Quetzalcoatl, a god opposed to Tezcatlipoca, and inhabited by humans who ate piñón nuts. This era was ended by a great windstorm on a day 4-wind. The third era, Tletonatiuh, was ruled by the rain god, Tlaloc, and was destroyed by a rain of fire on a day 4-rain. The fourth sun, Atonatiuh, was ruled by Chalchiuhtlicue, the female counterpart of Tlaloc; the population subsisted on a plant that was a precursor to corn. This era was destroyed on a day 4-water by a great deluge that made the sky fall and plunged the world into darkness (Nicholson 1971: 398–402).

The essential message conveyed by this myth is one of fatalism mixed with hope. The world would inevitably be destroyed but its destruction would give rise to a new creation. Life arose from death and set the stage for another cycle. Carrasco (1982: 99) pointed out that the collection of creation myths showed that creation took place through the coincidence of opposites—when dualities either struggled against or cooperated with each other. Each age was created by a pair in combat or sacrifice.

According to the Aztecs, we are now living in the fifth sun, which will be destroyed by earthquakes on a day 4-movement. After the destruction of the fourth era, the gods met at Teotihuacan to recreate the sun and to give life to the world. Two gods, Tecciztecatl, who was wealthy, and Nanahuatzin, the "Little Pimply One," were competing for the chance to jump into a fire and become the sun. Tecciztecatl claimed priority because of his wealth but balked three times. Nanahuatzin, however, did not hesitate to immolate himself and became the sun. Tecciztecatl, ashamed, followed him, but, because of his earlier cowardice, the other gods struck him and converted him into the moon. The sun was stationary, yet needed mobility to fulfill his creative role, so the other gods voluntarily drew their blood to nourish the sun and enable it to move (fig. 2.2). The myth conveys the symbolic message that the gods sacrificed themselves to create the world and its animal and plant life, and that thus mankind owes them a debt.

The self-sacrifice of the gods was not sufficient to guarantee the continued existence of the world; the sun had to be fed human blood daily in order to continue its daily round. Man's primary responsibility was feeding the sun, which warfare made possible through the sacrifice of the blood and hearts of captives (Nicholson 1971: 402). López Austin (1988: vol. 1, 74) claimed that the relationship between god and man was mercantilistic: man gave blood, hearts, and fire to the gods, and was given in return crops, water, and freedom from disease and plagues. This claim is supported etymologically. The word for sacrifice to the gods is *nextlahualiztli* ("the act of payment"), and the offering of fire to the gods, *tlenamaca*, means "to sell fire."

The myth about the creation of mankind embodies the theme of the gods' sacrifice for man. The Aztecs believed that Quetzalcoatl descended to the underworld to retrieve the bones of the inhabitants of the first creation from the ruler of the underworld, Mictlantecuhtli. After overcoming obstacles imposed by Mictlantecuhtli, Quetzalcoatl started to return to earth with the bones but

Figure 2.2.   Quetzalcoatl drawing blood to feed the Sun. (Sahagún 1906. *Florentine Codex*: book 3)

was chased by quail sent by Mictlantecuhtli, who wanted to re-nege on the deal. Quetzalcoatl fainted, dropping the bones, which the quail pecked at and broke. Quetzalcoatl wept, gathered up the pieces, returned to the earth, and gave the bones to the assembled gods, who ground them to a powder with a pestle. Quetzalcoatl pierced his penis and fertilized the powder with his dripping blood (Anonymous 1975: 120–121). After four days, a male child emerged from the vessel containing the pulverized bones, and a female followed four days later. All of mankind descended from this pair. León Portilla (1963: 111) felt that the etymology of the word for commoner, *macehualli*, reflects this myth. Man was the product of the gods' penance, so people were called *macehualtin* ("those deserved and brought back to life because of penance").[1] The message was clearly stated in a purported dialogue between Aztec priests and Spanish missionaries:

> It was the doctrine of the elders
> that there is life because of the gods;
> with their sacrifice, they gave us life.
> In what manner? When? Where?
> When there was still darkness. (León Portilla 1963: 64)

⸻ Man as Microcosm

Carrasco (1982: 63–103, 160–174), following Paul Wheatley, pointed out that Mesoamerican societies held "cosmo-magical" beliefs that presupposed intrinsic parallels in cosmic processes. He gave as an example the movements of the heavens in parallel with biological rhythms on earth—the annual cycles of plant regeneration and the functions of the human body.

In its layout of state capitals and other sacred precincts as microcosms, the architecture of the city or temple dramatized the cosmogony and made it tangible. Architectural symbolism, in turn, helped legitimize the status and power of the rulers: it underscored their identification with the gods and the power of the cosmos. The ideal city was a sacred space oriented about a particularly sacred temple or pyramid believed to be the center of the world, the *axis mundi*, where heaven, earth, and underworld met (Carrasco 1982: 71).

Recent excavations of the site of the main temple of Tenochtitlan (now downtown Mexico City) have shown that the temple itself is an architectural reenactment of the sun god, Huitzilopochtli, creation myth (Matos Moctezuma 1987: 57). The main temple of Tenochtitlan represented the hill Coatepec itself, as indicated by huge serpent heads signifying its name. The temple of the victorious Huitzilopochtli was placed at the top of the pyramid. A large circular relief sculpture of the dismembered moon goddess, Coyolxauhqui, and some anthropomorphic sculp-

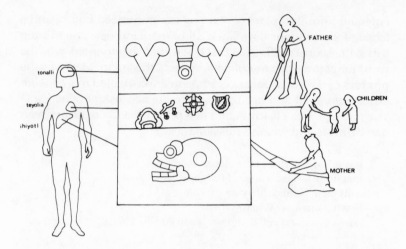

Figure 2.3.   Correspondence of animistic forces, levels of the universe, and the nuclear family. (Lopéz Austin 1988. *Human Body and Ideology*, by permission of Alfredo López Austin)

tures, which may represent her brothers, the Centzon Huitznahua, were placed at the bottom as a symbol of their defeat (Matos Moctezuma 1987: 57).

López Austin (1988: vol. 1, 161–164, 346–349) showed that metaphors were used to connect the human body to the universe. Corn was called *tonacayo* ("our flesh"), metaphorically linking humans and the cereal staple to which they owed their lives. The connection was made also in the myth that Quetzalcoatl and other gods traveled to the underworld in order to obtain corn for mankind. The connection was more direct in the Quiche Maya origin myth that God made man out of corn dough (Thompson 1966: 177). In rituals, the human body was called Chicomoztoc ("Seven Caves"), metaphorically equating body orifices with the mythical place of origin of the Aztec tribes before their migration to Central Mexico (Ruiz de Alarcón 1984: 176).

In addition to the twofold division of the universe, the analogy between the cosmos and the human body included a tripartite correspondence. The cosmos was divided into three levels—the upper heaven (*ilhuicatl*), the lower heaven, and the underworld—linked to the three animistic forces (or souls) residing in the body (fig. 2.3): ilhuicatl to the head and the animistic force *tonalli*, the lower heaven to the heart and the animistic force *teyolia,* and the underworld to the liver and the animistic force *ihiyotl.* Dualism associated ilhuicatl with the father and masculinity and the underworld with the mother and femininity, while the children occupied the liminal position of the lower heaven (López Austin 1988: vol. 1, 348). These animistic forces will be detailed below.

A linguistic marker for the cosmo-anatomic analogy can be seen in the word for cultivating the earth with a digging stick—*elimiqui* ("to harm the liver"); the earth was being wounded by the farmer's labor. The interrelationship between man and nature was noted by the use of the same nomenclature for parts of both the human body and trees. Thus, *cuauhtlactli* ("tree trunk") contains the root *tlac* also used to refer to a human torso. Similarly, tree branches are "hands," the top of the tree is "hair," the bark is "skin," and the wood is "flesh." The social order of human society also is paralleled in nature. There are *tecuhtli* ("rulers") of birds, of serpents, and even of ocelots. As we mentioned, the center was associated with rulership, so there was a "heart" of the mountains, of the lakes, and of the earth (López Austin 1988: vol. 1, 347). This anthropomorphic universe linked terrestrial and celestial processes with the biological processes of the human body. For example, the movements of celestial bodies influenced the fate and health of humans; reciprocally, the tears of children to be sacrificed during Tlaloc's festivals produced rain.

= Complexes of Deities

The Aztec pantheon was large and confusing. Deities were anthropomorphic and animistic; almost any aspect of nature had a deity associated with it. Many deities had animal alter egos called *nahualli.* Gods had dual and quadruple natures, the former associated with the concept of opposition and the latter with the cardinal directions. Particular sociopolitical groups (towns, trades, professions, and social classes) had specific tutelary deities.

The Aztec tendency to incorporate the gods of conquered peoples into their own pantheon caused further complications. Nicholson (1971) attempted to produce some order by identifying three major themes expressed by these deities: 1) celestial creativity-divine paternalism, 2) rain-moisture-agricultural fertility, and 3) war-sacrifice-sanguinary nourishment of the sun and earth. Gods may then be grouped into deity complexes under these major themes. A deity complex is composed of a number of deities closely related in function and iconography. They may be variants of a basic deity, because they are its nahualli, or its double, because they are its geographical variations. The following brief list is illustrative (Nicholson should be consulted for a more extensive discussion of the god complexes and a more complete list of the gods):

1. Celestial Creativity-Divine Paternalism
    Ometecuhtli complex (God of Duality)
    Tezcatlipoca complex (Creator God)
    Xiuhtecuhtli complex (Fire God)
2. Rain-Moisture-Agricultural Fertility
    Tlaloc complex (Rain God, figs. 2.4, 2.5)
    Centeotl complex (Corn God)
    Xochipilli complex (God of Spring)
    Ometochtli complex (agricultural fertility)
    Teteo Innan complex (Earth Goddess)
    Xipe Totec complex (fertility and Spring)

Figure 2.4.   The Aztec Rain God, Tlaloc, a leading member of the Rain-Moisture-Agricultural Fertility Complex. (*Codex Laud* 1964: fol. 12)

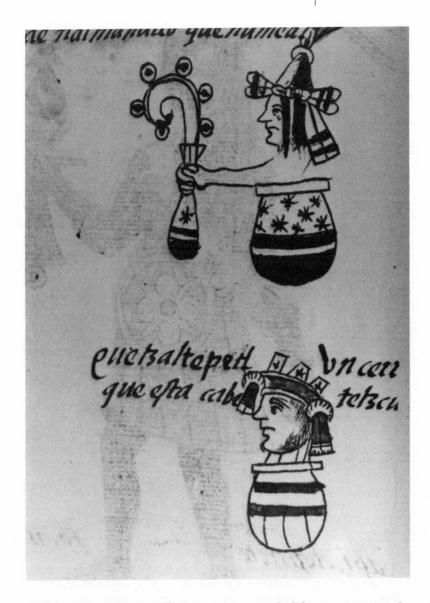

Figure 2.5. Dwarfs called *tepictoton* and *tlaloque,* associated with and assistants of Tlaloc. Some were the *teyolia* ("souls") of people selectively killed by Tlaloc. (Sahagún 1906. *Florentine Codex:* book 1)

3. War-Sacrifice-Sanguinary Nourishment of Sun and Earth
Tonatiuh complex (Sun God)
Huitzilopochtli complex ("Hummingbird on the Left"—solar patron god of the Mexica)
4. Quetzalcoatl—a very complex deity who combined the characteristics of several themes. He was a prime Creator God but also participated in the Rain-Fertility complex.

Gods also had important functions as patrons of the days and of longer divisions of the calendar. In this capacity they were involved with the astrological aspects of health and illness. Violations of ritual or offenses against a particular god could result in specific ailments, to be discussed later.

= *Preservation of the Social Order through Social Sanctions*

Wigberto Jiménez Moreno (1971) pointed out a fundamental distinction between Mesoamerican religions and Christianity:

> There is a fundamental distinction between indigenous religions and Christianity. In the latter, the major emphasis is put on the salvation of the soul, and thus on the welfare of the individual in the after-life. While in pre-Hispanic paganism, the emphasis is put on preserving the cosmic order, and the individual, as such, has almost no value in isolation except to the extent that he contributes to collective activities that have the conservation of that order as a goal. While a Christian thinks that one must perfect one's character and must love others, in indigenous religions the ethical norms put greater emphasis on the good of the group rather than that of the individual.

Aztec rulers identified themselves with the central axis of the universe, the power associated with it, and with creation itself, setting themselves off symbolically as a race apart from ordinary human beings (Coe 1981: 183; Carrasco 1982: 151, 161). Duty to the rulers and to the State became a religious command. Warfare, particularly for the purpose of capturing prospective victims to feed the sun, became a sacred mission. Tlacaelel, the most famous Cihuacoatl of the Mexica, is credited with promoting the idea that the Mexica were the "People of the Sun" entrusted by God to ensure that the world would not end due to a lack of human sacrifice (Durán 1967: vol. 2, 232–234). Aztec origin myths served to ideologically justify not only human sacrifice but also, for example, ideological extension of the Aztec empire by conquest. They were also the basis of social control mechanisms, of the praise of battlefield courage, and of popular support for the militaristic ends of the group in power. They also justified the sumptuary privileges of a nobility based on military prowess and valor. A large component of social control involved the etiology of diseases, whose cause-effect relationships could be explained only by the supernatural (López Austin 1988: vol. 1, 244).

The myths considered so far have to do with justification of human sacrifice from the viewpoint of the sacrificer. However, religions have to be internally consistent; a converse set of beliefs related to the perspective of the victim. A person's fate after life did not depend primarily on earthly behavior but on the manner of death. Persons chosen by Tlaloc to die, for example, those who drowned, were hit by lightning, or died of a disease associated with Tlaloc, went to Tlalocan, a place of abundance and joy. However, one could not choose to die in this fashion. People who died ordinary deaths by illness or accident went to Mictlan, the place of the dead. The voyage took four years, involved a series of trials and hazardous locations, and apparently resulted in oblivion. The best death was in battle or, its equivalent, as a sacrificial victim. Such lucky souls were destined to accompany the sun in its travels from east to west from sunrise to noon. After four years, the souls of these warriors/victims were reborn as hummingbirds or butterflies. These beliefs were a powerful rationale for a willingness to die in battle or sacrifice. The following is one of many poems exalting this philosophy (León Portilla 1969: 86–87): [2]

> From where the eagles are resting,
> from where the tigers are exalted,
> the Sun is invoked.
>
> Like a shield that descends,
> so does the Sun set.
> In Mexico night is falling,
> war rages on all sides.
> Oh Giver of Life!
> War comes near. . . .
> Proud of itself
> is the city of Mexico-Tenochtitlan.

Here no one fears to die in war.
This is our glory.
This is Your Command,
oh Giver of Life!
Have this in mind, oh princes,
do not forget it.
Who could conquer Tenochtitlan?
Who could shake the foundation of heaven?

With our arrows,
with our shields,
the city exists.
Mexico-Tenochtitlan remains.

Childbirth was considered symbolic combat. A woman in labor would be given a toy shield and spears to fight her battle, and midwives would utter a war cry, which according to the text meant that "the woman had fought her battle well, that she had been a valiant warrior, that she had taken a captive, that she had captured a child" (Sahagún 1950–1969: book 6, 167). Women who died in the battle of childbirth accompanied the sun in its journey from noon to sunset. Such women, the *cihuateteo* ("women goddesses"), were not reborn as hummingbirds but instead returned to earth on certain days associated astrologically with the west (fig. 2.6). Anyone, particularly a child, who encountered them, would be stricken with a fit resembling epilepsy (Sahagún 1950–1969: book 1, 19).

The Aztecs were very concerned with language. Words, in themselves, had magical power to harm or heal. Oratorical performance was highly regarded. Rulers and nobles were as proud of composing poems as of any other accomplishment, because those who excelled in rhetoric were *yolteotl* ("deified heart" infused by divine wisdom). This interest in words extended to the Aztec belief that ultimate truth or real knowledge was not achievable by empirical observation, but only through the medium of poetry, *in xochitl in cuicatl* ("flower and song") (León Portilla 1963: 73–79). Because of this, poetry, various two-word metaphors characteristic of Aztec rhetoric, and *huehuetlatolli* ("talk of the elders," that is, admonitory discourses) are important sources for Aztec views of the nature of life and man.

Huehuetlatolli are often couched as admonitions and instructions concerning the proper and ideal forms of behavior and thought. They must be given weight as statements of the societal

Figure 2.6. The Cihuateteo, *teyolia* of women who had died in childbirth, descending to earth. (Sahagún 1906. *Florentine Codex*: books 4/5)

ideals advocated by the ruling class. For example, a huehuetlatolli directed at a newborn said:

> Thou wert sent here on earth. Thou camest not to rejoice, thou camest not to be content; thou camest that thy bones, thy body should endure pain, suffer affliction. And thou wilt work like a slave, thou wilt labor, thou wilt suffer weariness here on earth. For this reason wert thou sent. (Sahagún 1950–1969: book 6, 183–184)

A huehuetlatolli directed at seven-year-old girls advised:

> There is no rejoicing, there is no contentment; there is torment, there is pain, there is fatigue, there is want; torment, pain dominate. Difficult is the world, a place where one is caused to weep, a place where one is caused pain. . . . And it is a place of thirst, it is a place of hunger. This is the way things are. . . . it [the earth] is not a place of contentment. It is merely said it is a place of joy with fatigue, of joy with pain on earth. . . . In order that we may not go weeping forever, may not die of sorrow, it is our merit that our lord gave us laughter, sleep, and our sustenance, our strength, our force, and also carnal knowledge in order that there be peopling. (Sahagún 1950–1969: book 6, 93)

A huehuetlatolli at a funeral said:

> Oh my son, thou hast found thy breath; thou hast suffered; our lord hath been merciful to thee. Truly our common abode is not here on earth. It is only for a little time, only for a moment that we have been warm. Only through the grace of our lord have we come to know ourselves. (Sahagún 1950–1969: book 3, 41)

The concept of an ephemeral life on earth was repeated in a poem by the ruler Nezahualcoyotl:

> Is it true that on earth one lives?
> Not forever on earth, only a little while.
> Though jade it may be, it breaks;
> though gold it may be, it is crushed;
> though it be quetzal plumes, it shall not last.
> Not forever on earth, only a little while. (León Portilla 1963: 72)

Life was seen as an ephemeral moment, and earth as a place where suffering, metaphorically expressed as fatigue or physical pain was the norm. Man was destined to feel hunger and thirst. The proper attitude was forbearance, stoicism, and the faithful performance of one's work and duty. Aztec society was stratified, with large differences in wealth and living standards. The common man, the macehualli, led a life of intense labor, with some bouts of hunger and the possibility of premature death through combat or sacrifice. One of the functions of the ideology exalting stoicism and duty was to reconcile people to their lot by attributing suffering to the immutable will of the gods and to the nature of life on earth itself rather than to the existing social structure. As the huehuetlatolli mentions, there is laughter, joy, and sex in this world, but these are ephemeral, and even sex is connected to duty—the peopling of the earth.

Social stratification was justified as due to actual physical and spiritual differences reflecting the superiority of nobles. Although the origin myth states that woman took four more days than man to emerge from the powdered bones of previous generations, classical sources do not specifically speak of the superiority of man. However, the Tzotzil Indians, among others, believe that men are born with a stronger tonalli than women (Guiteras Holmes 1965: 249). Nobles were born with a stronger tonalli than commoners and preserved it better by adhering to a higher moral standard and by remaining chaste longer. The holding of governmental posts and the fulfilling of official duties also strengthened their tonalli and reciprocally justified their elevated status.

These ideas were echoed in the two-word metaphors used to describe the classes. Commoners were *cuitlapilli atlapalli* ("tail and wing"). If the state was a bird, the common people were the motive power but not the head. Commoners were like children who were not self-sufficient and must be cared for. They were *in itaconi in mamaloni* ("that which is carried, that should be shouldered"), *in quimilli in cacaxtli* ("the bundle, the packframe"), or *in tecuexanco in tememeloazco* ("that which goes in one's lap, cradled in one's arms"). A huehuetlatolli refers to the commoners as "the poor, he who is carried, who is shouldered, the tail, the wing, he looks for his mother, his father, he wants to be ruled" (Sahagún 1950–1969: book 6, 23). In contrast, the rulers were referred to as *motenan motzacuil* ("your rampart, your refuge") or *in ahuehuetl in pochotl* ("the bald cypress, the ceiba," both towering trees), that is, images of strength and support.

Nobles also received divine sanction to rule. According to myth, at the time of the Fifth Creation, nobles (pipiltin), who were the descendants of Quetzalcoatl, were created to be rulers (Sahagún 1950–1969: book 6, 83).

The *tlatlacoliztli,* a temporary slavelike status, also was characterized by a physical difference. Castillo (1972: 119) relates the etymology to either *tlaco* ("a half") or the verb *tlacoa* ("to sin, to spoil, to harm"). Indications suggest that what was harmed or reduced was the tonalli or the teyolia. Because the tonalli was associated with hair, particularly that on the crown of the head, shaving it and thus diminishing the tonalli would turn a person into a *tlacotli,* or slave (López Austin 1988: vol. 1, 402). As we have seen, the loss of tonalli could precipitate a fever, which in this case might overheat the heart. It was said that the son of a slave "was born in the incense ladle," while a free man was called a *yolloiztic* ("cold heart") or *yolloxoxouqui* ("green or uncooked heart"). This condition was reversible. A slave could be freed through a rite of passage, which (inexplicably) involved stepping on human excrement and ritual bathing to restore what had been lost (Durán 1967: vol. 1, 185).

López Austin (1988: vol. 1, 185–190) pointed out that the word *tlacatl* ("man, human") is also related to the tlatlacoliztli concept because its literal meaning is "diminished man." No Aztec myth explains this, but a Quiche Maya myth in the Popol Vuh is relevant. In this myth the first humans created were so intelligent and had such good vision that the gods grew jealous of them. The gods blew a mist into their eyes and thus "diminished" their vision and wisdom to the present level (Recinos, Goetz, and Morley 1950: 167–169). Similarly, living humans were shorter than the first humans, who were giants, as a result of Quetzalcoatl's fall during his return from Mictlan and the consequent breakage of the giants' bones he was carrying.

Another important buttress of the social order and of duty as an ethic was a strong emphasis on moderation, as seen in the huehuetlatolli:

> On earth we live, we travel along a mountain peak. Over here there is an abyss, over there is an abyss. If thou goest over here, or if thou goest over there, thou wilt fall in. Only in the middle doth one go, doth one live. (Sahagún 1950–1969: book 6, 101)

> They went saying that on earth we travel, we live along a mountain peak. Over here there is an abyss, over there is an abyss. Wherever thou art to deviate, wherever thou art to go astray, there wilt thou fall, there wilt thou plunge into the deep. That is

to say, it is necessary that thou always act with discretion in that which is done, which is said, which is seen, which is heard, which is thought, etc. (Sahagún 1950–1969: book 6, 125)

The basic rules were equilibrium, moderation, and performance of duty. States of health or illness were closely related to a condition of equilibrium or disequilibrium, respectively. For health one had to be in equilibrium physically, in one's social relations, and with the deities. Social control was accomplished by holding the body hostage and punishing deviations with illness. Punishment often did not require active intervention by sorcerers, spirits, or gods because it followed mechanically and automatically from the properties and characteristics of animistic forces (described below). Moderation in diet, exercise, and behavior was an essential component of a balanced body. Work and tiredness created disequilibrium, among other ways, by overheating a person's tonalli. Indians still believe that sin, states of anger, getting overtired, or sexual excesses make people vulnerable and can lead to illness.

═ *Multiple Souls (Animistic Forces): Tonalli, Teyolia, and Ihiyotl*

The Aztecs believed that the human body had several animistic forces (souls), each with specific functions for the body's growth, development, physiology, and even its fate after death. The three principal animistic forces were: tonalli, located in the head; teyolia, in the heart; and ihiyotl, in the liver (fig. 2.7). The health of an individual depended on the relative amounts of each soul at a given time and on the maintenance of a balance among them.

The word tonalli comes from the root *tona*, meaning heat, which is also found in Tonatiuh, the Sun God. Tonalli has various meanings: irradiation, solar heat, astrological sign of the day, fate of a person, soul and spirit, thing that is destined for or the property of a given person. All these connotations appeared in the intimate connections between tonalli and the complex Mesoamerican calendar. The calendar was based on the interaction between a 365-day solar calendar (divided into eighteen months of twenty days and one "month" of five days) and a sacred calendar, *tonalpohualli* ("count of the tona"), which lasted 260 days and was

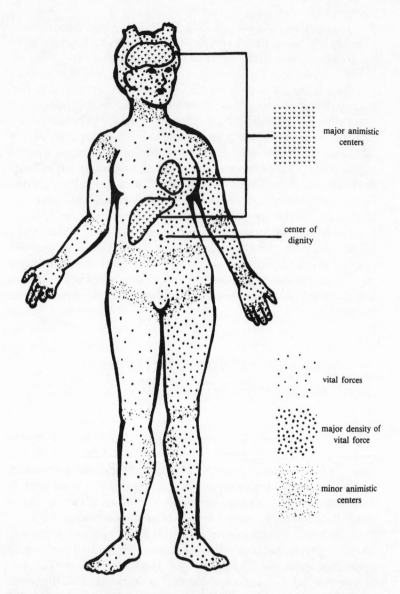

Figure 2.7.   Location of the animistic forces in the body. (Lopéz Austin 1988. *Human Body and Ideology*, by permission of Alfredo López Austin)

elaborated by combining twenty day names and the numbers one through thirteen. The sacred calendar and the patron deities of the day determined the fate of persons as well as the omens for performing specific activities. Specialists, the *tonalpouhque* ("enumerators of the tonalli"), were consulted to learn the outcome of the complex interplay of the various astrological influences on a particular day. They predicted the fate of a newborn child and selected propitious dates for agricultural or commercial activities.

The date of birth determined a person's occupation. For example, people born on a day 4-dog would do well at raising dogs, while women born on a day 1-flower would become weavers and seamstresses. It could also determine a person's life span and general health. People born on 1-twisted grass would have many children but all would die young (Sahagún 1950–1969: book 4, 55). Those born on a day 6-dog would be sickly and weak and would soon fall dead—"If he lived, he would live only in suffering . . . would go about constantly and continually coughing, pallid, green-faced, white with cold, etc." (Sahagún 1950–1969: book 4, 73). In contrast, those born on 1-lizard would be very strong and healthy. They, like lizards, would even be able to survive falls without harm (Sahagún 1950–1969: book 4, 83).

Fates were not immutable. Tonalpouhque priests could ordain a shift in a child's baptismal day to make it appear as if the child had been born on a day with a better prognosis. Many forecasts in the *tonalamatl* ("book of the days or fates") were presented in a conditional fashion. Apparently, the tonalli was not a fixed quality but rather a temperament or a propensity to behave in certain ways, which could interact with the other animistic forces, the social environment, or life experiences to produce a final result. Thus, different fates were possible depending on both an individual's tonalli and particular circumstances. Failure to be properly humble, to perform one's duty, or to honor the gods could nullify a good omen. Conversely, diligent prayer and penance could ameliorate a poor one. For example, a man born on a day 1-flower was destined to be a skilled artisan and wealthy. However, if he became too proud and was not respectful to his elders, the gods would punish him with great illnesses and cause him to die in poverty (Sahagún 1950–1969: book 4, 23–24). Women born on a day 7-flower would normally attain fame and fortune as embroiderers, but if they neglected to do the fasting required to honor their patroness, Xochiquetzal, they would become harlots (Sahagún 1950–1969: book 4, 7). Men born on 1-ocelot were destined to be captured in war and sold into slavery or become adulterers. However, if they diligently performed penances to the gods and were conscientious in carrying out their duties, their fate could possibly be altered so they could achieve a proper life (Sahagún 1950–1969: book 4, 5–6).

From the moment of conception, tonalli linked a person to the gods and to the universe. It was infused into the fetus, before implantation in the womb, by Ometecuhtli, the Creator God of Duality. Tonalli was a force that gave the individual bravery, vigor, and warmth and made growth possible. Fetal tonalli alone was not vigorous enough to enable the newborn child to prosper, so rituals were performed to strengthen the child's tonalli. The sun, Tonatiuh, was the best source of tonalli. Meanwhile, an alternate form of heat was needed, because it was thought to be dangerous to expose a newborn to the sun before determining what the omens were for its birthday. A fire was lit in the birth room and was kept burning constantly for four days until the baby was ritually bathed and named by the tonalpouhque. The number four was closely related to fire and to the sun (Caso 1967: 190–199). Ometecuhtli was asked to give the child additional tonalli (López Austin 1988: vol. 1, 209–214).

The link that tonalli made between an individual and the heavens was apparently regarded as not only symbolic but physical in nature, perhaps similar to a cord, but invisible. Evidence for this is the belief that someone jumping over a child who was lying down would harm the child's tonalli and stunt its growth. To undo the harm, it was necessary to jump over the child in the reverse direction (Sahagún 1950–1969: book 4, 184). Similarly, to step over someone's head was considered a great insult, because protecting one's head was equivalent to protecting one's reputation and fate.

The amount of tonalli varied with age, circumstance, and social position. Beliefs concerning tonalli reinforced and upheld the predominant ethic: the primary good was service to the State. Because tonalli increased with age, very old and respected people (mostly males) had very strong tonalli. They were said to be *tleyotl*, literally "full of fire" but denoting fame and honor (López Austin 1988: vol. 1, 257). Tonalli could also be increased by performing important duties for the State, a physiological reward reinforcing the ethic of duty. Nobles (*pilli*-singular; *pipiltin*-plural) also had more tonalli than commoners, presumably because they needed more vigor to be able to perform the arduous duties of governing. Good performance of these duties in turn increased their tonalli. This served to doubly validate the right of nobles to rule: they merited authority by both meeting a higher moral standard and having physically superior tonalli.

Differences in the amount of tonalli arose from several sources. Pipiltin were born with a stronger god-given tonalli (the tonalamatl often gives different omens for nobles and commoners born on the same day), a congenital difference that could be augmented during life by bravery on the battlefield, performing well in governmental positions, and adherence to a higher moral standard. For example, since premature or excessive sexual activity could diminish tonalli, and hence growth and intelligence, noble youths were expected to remain chaste longer than youths of common birth. Rulers, who had a special need for tonalli, could also increase it through other means. Slaves were sacrificed to add vitality to the ruler and to extend his life (Sahagún 1956: vol. 1, 334–335). The tonalli of rulers could also be strengthened with the brains, blood, and bile of wild beasts, and with the fragrance of flowers (de la Cruz 1964: 193; Hernández 1959: vol. 1, 389–390).

Tonalli could be decreased by becoming drunk. Consequently, the overuse of *octli* (*pulque*—fermented agave sap) was regarded as a danger to both health and morals. Drunkenness implied a loss of control, a vulnerability to spirit possession, and an inability to perform one's duty, the greatest crime of all. Nobles or priests found drunk were punished by death. However, drunken commoners merely had their heads shaved for the first offense, and only repeat offenders were executed. The connection between punishment and failure in duty is evident in the fact that old men and women, who had already discharged their duty to the State, were allowed to become drunk with impunity (Soustelle 1970: 157).

As shown in chapter 4, octli was regarded as an important food source, so complete abstinence was not indicated. Rather, moderation, as in all things, was the proper stance. The effects of drinking octli, as well as of ingesting hallucinogens, were attributed to possession by spirits that inhabited these substances. In the case of octli, the spirits were the Tzentzon Totochtin ("400 Rabbits"), agricultural fertility gods and members of the Ometochtli ("God Two Rabbit") complex. The number 400 was used as a metaphor for "many" and explained the variety of drunken behaviors (pugnacious, sleepy, weepy, jovial, morose, etc.) by attributing different types to possession by different "rabbits" (López Austin 1988: vol. 1, 355–356). People born on a day 2-rabbit were destined to be drunks. To get drunk was to "rabbit yourself."

Tonalli could leave the body only for brief intervals because it was essential for life. Distinctions were made between voluntary and involuntary as well as between normal and abnormal departures. López Austin (1988: vol. 1, 221–222) states that, among modern Mexican Indians, *sombra* ("shadow") is the equivalent of tonalli. Sombra can leave the body normally and involuntarily when one is drunk, unconscious, asleep, or engaged in sexual intercourse. López Austin found indications that classical Aztecs held the same beliefs. Since tonalli left the body during intercourse, admonitions were given against a sudden interruption because the tonalli might be unable to return to the body, possibly resulting in illness. Dreams were believed to be the perception by a person's tonalli of real locations away from the body. These beliefs are consistent with the strong shamanic component in Aztec religion. Shamanic religions believe in the separability of the soul from the body. Shamans differ from other people in that they can voluntarily send their souls, in this case their tonalli, on out-of-body excursions without being harmed. Involuntary loss of tonalli caused illness and death. These diseases will be discussed more fully in chapter 6.

López Austin (1988: vol. 1, 216–217), indicating the derivation of the word tleyotl (fame, honor) from the root for fire, believed that it was also used as a synonym for tonalli and that therefore the nature of tonalli was to be hot. Tonalli supplied heat to the body but also somehow functioned as a thermostat, a loss of tonalli apparently resulting in a fever (López Austin 1988: vol. 1, 216). We will deal at length later with the question of whether the Aztecs had a concept that an imbalance of "hot and cold" caused illness before the Spanish conquest. Here we give a couple of examples involving the Aztec view of the effect of "heat" and "cold" on the human body and cite aspects that would have been alien to a European mind, explainable only by the Aztec concepts of tonalli. One example is the Aztec beliefs concerning twins. It was said that if twins were nearby, steam baths would not heat properly, food would not cook well, and the color red would not dye evenly. López Austin (1988: vol. 1, 256–257) pointed out that Indians today consider twins to be "cold" and assumes that this was also true in the past. This lack of heat must have been attributed to the division of tonalli between two beings in the same womb, so that each received less than a normal amount. Twins would therefore suck heat away from sources such as steam baths or cooking fires. In a second example, this concept of tonalli sharing may have been the Aztec's proposed etiology of an illness known as *tzipitl*, affecting small children whose mothers became pregnant again too soon. Its symptoms were weakness, diarrhea, slow growth, and loss of weight. Linguistically, it was seen as an illness due to a tie between the nursing child and the fetus, similar to a sharing of tonalli (López Austin 1988: vol. 1, 259). Now we would say that the child was suffering from kwashiorkor, a protein-calorie deficiency syndrome due to premature weaning or otherwise inadequate diet.

Teyolia ("that which gives life to people"), the second animistic force, resided primarily in the heart. It imparted vitality, knowledge, and vocational ability to its possessor because the heart was the center of thought and personality, involved in will, memory, emotion, and mental activity. For example, people who had achieved distinction in the arts, poetry, or intellectual pursuits were called yolteotl ("deified heart") because their hearts had received a divine force (López Austin 1988: vol. 1, 231–232). Teyolia was infused into the fetus and reinforced at ritual rebirth (the bathing and naming ceremony) when Chalchiuhtlicue was asked to give additional teyolia to the child (López Austin 1988: vol. 1, 209–214). This soul, unlike tonalli, was inseparable from the human body and was identified as the soul "that goes beyond after death." Damage to the heart and the teyolia could occur in various ways. A group of diseases (see chap. 6) that caused people to go crazy or to have fits were thought to be caused by an excess of phlegm in the chest and/or an intrusion by the teyolia of Tlaloc's helpers.

The heart could also be harmed by immoral conduct. Tlazolteotl, the goddess of sexual lust, tempted humans to carnal excesses that damaged and distorted the heart. A huehuetlatolli to young men exalts the theme of sexual moderation and cautions against the physical harm that results from excess:

> Do not throw yourself upon women
> like the dog which throws itself upon food.
> Be not like the dog
> when he is given food or drink,
> giving yourself to women before the time comes.
> Even though you may long for women,
> hold back, hold back with your heart
> until you are a grown man, strong and robust.
> Look at the maguey plant.
> If it is opened before it has grown
> and its liquid taken out,
> it has no substance.

It does not produce liquid; it is useless.
Before it is opened
to withdraw its water,
it should be allowed to grow and attain full size.
Then its sweet water is removed
all in good time.

This is how you must act:
before you know woman
you must grow and be a complete man.
And then you will be ready for marriage;
you will beget children of good stature,
health, agile, and comely. (León Portilla 1963: 149–150)

Following the duality principle, the cure for the disease of sexual excess was confession, in the presence of a priest, to Tlazolteotl, the agent of both temptation and salvation. This act of penance and confession called *neyolmelahualiztli* ("the act of straightening out hearts"), could be performed only once in a lifetime.

Moderation was particularly important in sexual matters. Premature sex was thought to stunt growth and intelligence by loss of tonalli. Aztecs apparently believed that men possessed a fixed amount of semen because the later a man waited to start sexual activity the later in life he could still perform. Women were potentially insatiable, as was pointed out in a dialogue between Nezahualcoyotl ("hungry coyote," a famous ruler of Texcoco) and two elderly women caught committing adultery with younger men. Nezahualcoyotl asked them why they still desired sex at such an old age, and they replied:

You men cease to desire sexual pleasure when you are old because you indulged in your youth and because human potency and semen run out. We women, however, are never satiated nor do we stop liking it. Our body is like a cave, a gorge which never swells and receives whatever is thrown in and still wants and demands more. And if we don't do this we are not alive. I tell you this, my son, so that you will live cautiously and discreetly; so you will go step by step and not hurry with this ugly and harmful business. (Sahagún 1956: vol. 2, 146)

Sexual pleasure was allowed for adults, if only "to people the earth," but only in moderation.

Excessive sex at any age was dangerous. It would "fill your tonalli with filth" and lead to illnesses characterized by emaciation, coughing, a blackened body, and pus in the urethra (López Austin 1971b: 143, 201). In keeping with the Aztec ideal of moderation, it was also harmful not to engage in sex when appropriate. Widowers were supposed to eat ocelot meat to cool their bodies in need of sex, and thus prevent illness (López Austin 1971b: 213). Celibacy, too, was condemned, and rape was punished by death. Active or passive homosexual acts by either sex were punished by death. Engaging in such illicit sexual relationships was harmful in two ways; homosexual acts harmed the participants themselves, who subsequently emitted ihiyotl, which could harm others (López Austin 1988: vol. 1, 304, 307). Sterility, attributed mostly to the woman, could be used by her husband as grounds for divorce. The large number of remedies for sterility found in the sources indicate that it must have been common (Quezada 1975a).

The third animistic force, ihiyotl ("breath, respiration") was concentrated in the liver. It gave humans vigor, passions, and feelings such as desire, envy, and anger. It was thought to be a luminous gas with the property of attracting other beings. Etymological evidence for this is the name for a magnetic stone—*tlaihiyoanani tetl* ("stone that attracts things with ihiyotl"). Several expressions about the liver give an idea of the range of functions served by ihiyotl. To act with the liver, *eltiliztli*, was to be careful or to make love to someone. To be lazy was to "lack a liver" while to have a swollen liver was to be angry. A person whose interior forces were in harmony had a *cemelli* (a united liver) and was happy and full of pleasure. A virtuous person had a clean liver, while a dirty liver was associated with immorality, particularly sexual transgressions. This attitude toward sex is reflected in the two-worded metaphor for sex—*in teuhtli in tlazolli* ("the dust, the garbage"); sexual transgressions covered the tonalli with dust or filled it with garbage (López Austin 1988: vol. 1, 222, 313–314). Another metaphor for carnal sins was *in cuitlatitlan in tlazoltitlan* ("in the excrement in the refuse") (Sahagún 1950–1969: book 6, 97). Sins, particularly sexual sins, harmed the liver and made it emit ihiyotl involuntarily, the invisible emanations causing a number of "filth" diseases" (see chap. 6).

An interesting illustration in the *Codex Laud* (one of the few surviving pre-Columbian codices) seems to show the disaggregation of the animistic forces at the time of death (fig. 2.8). Four serpentlike figures emerge from the body; two have serpent heads, one the mark of the god Ehecatl (an avatar of Quetzalcoatl), and one a skull. This codex has not been thoroughly studied, but

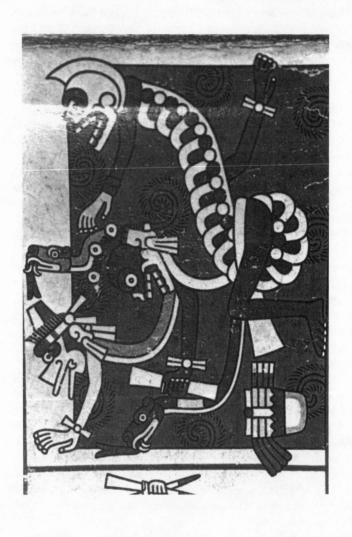

Figure 2.8. Animistic forces leaving the body after death. (*Codex Laud* 1964:fol. 21D)

López Austin (1988: vol. 1, 316) suggested the following: The serpent emerging from the head is the tonalli; the serpent emerging from the breast with Ehecatl's head is the teyolia; the serpent-headed figure emerging from the belly may be the ihiyotl; and the bare skull represents the empty shell left after departure of the animistic forces.

We have much more information about the fate of teyolia after death than about the other two animistic forces. Since teyolia was the "soul" that "went beyond" after death, the teyolia of warriors dying in combat or sacrificed accompanied the sun, and the warriors were reincarnated as hummingbirds. The teyolia of women who died in childbirth accompanied the sun until sunset, and the women became the cihuateteo. The women were not reincarnated; through their teyolia, they returned to earth on days associated with the west and afflicted humans who chanced to meet them. The teyolia of people chosen to die by the gods became the god's servants. The teyolia of those chosen to die by Tlaloc were called *ahuaque* or *ehecatotontin*, lived in caves in mountains, and assisted members of the Tlaloc complex in sending lifegiving rain but also harmful rains, storms, lightning, and hail to punish sinful towns. Such teyolia also participated in producing diseases associated with Tlaloc. The treatment of Tlaloc's chosen was unusual: their bodies were buried rather than cremated, although cremation was the standard procedure for transmitting the teyolia in smoke to the beyond.

People who died ordinary deaths were wrapped into a mummy bundle and cremated with items they would need on their long journey to Mictlan. The fate of those who went to Mictlan was mysterious and uncertain, even to the Aztecs:

> Given over to sadness
> we remain here on earth.
> Where is the road
> that leads to the Region of the Dead,
> the place of our downfall,
> the country of the fleshless?
>
> Is it true perhaps that one lives
> there, where we all go?
> Does your heart believe this?
> He hides us
> in a chest, in a coffer,
> the Giver of Life,
> He who shrouds people in the grave.

Will I be able to look upon,
able to see perhaps, the face
of my mother, of my father?
Will they loan me
a few songs, a few words?
I will have to go down there;
nothing do I expect.
They leave us,
given over to sadness. (León Portilla 1969: 85)

Little is known about the teyolia called *mictecah*, residing in Mictlan, except that they served as messengers of the rulers of the underworld (López Austin 1988: vol. 1, 339). Very possibly these teyolia were arthropods such as scorpions and spiders, considered very bad omens and forecasters of disease (Serna 1953: 225). Mendieta (1971: 97), in a somewhat garbled version, supported this hypothesis:

The people of Tlaxcala believed that the souls of lords and heads of state [those who died in warfare] became mists, clouds, birds of fine feather, and other things, and also precious stones of great worth. And that the souls of common people [those who went to Mictlan] became weasels, stinking beetles, animals that emit a stinking urine, and other thieving animals.

The final group of teyolia consisted of suckling infants who died. They had not become subject to the usual rules because 1) they had not eaten corn, that is, had not partaken of human food, and 2) had not engaged in sex. Thus their death was only a return to the heavens to await another opportunity to return to the earth (López Austin 1988: vol. 1, 314). These teyolia would go to a place called Chichiualcuauhco, a tree with leaves shaped like human breasts dribbling milk to the children's teyolia laying around the base of the tree (fig. 2.9).

Much less is known about the fate of tonalli after death. Tonalli's close association with hair was reflected in the symbol for battlefield capture—grasping the forelock of the captive—and in its relation to the ailment of fallen fontanelle (see chap. 6). Also, prior to cremation, a lock of hair cut from the top of the head was placed in a box with a similar lock cut from the same person shortly after birth. The hair was supposed to contain tonalli and act like a magnet to attract the remaining tonalli of the deceased wherever it was to be found. After cremation, the box containing the hair received the ashes and was topped with an image of the

Figure 2.9.   Chichiualcuauhco ("the Place of the Nursemaid Tree") where, according to the Aztecs, babies who died were transported and remained after death. (*Codex Vaticanus Latinus* 3738 1964: plate 4)

deceased. Ceremonies and offerings honoring the deceased were conducted for four days and yearly thereafter (Torquemada 1975–1983: vol. 4, 299–301; López Austin 1988: vol. 1, 322).

The fate of ihiyotl after death is not clear. Mexican Indians to-day believe that *aire de noche* ("night air"), the equivalent of ihiyotl, becomes a phantom that roams at night and can harm people (López Austin 1988: vol. 1, 324, 340–342). One of Sahagún's early versions mentions an illness called *yohualehecatl* ("night air") cured by cutting the skin with a flint, still the present-day treatment. This may be evidence that ancient beliefs about ihiyotl have echoes in current Indian culture (López Austin 1972: 138–139).

= Shamanism and the Aztecs

Aztec religion was unique in its combination of an elaborate state religion with shamanism, generally considered more

individualistic and less organized. Among the concepts and practices characteristic of shamanistic religions are: all phenomena in
the environment are animate; the soul is separable from the body
during life, susceptible to loss or to straying during sleep; a shaman in an ecstatic state can project his soul; the experience of an
altered state of consciousness as part of a shaman's initiation to
his calling; a "sickness" vocation into which a shaman is recruited
by surviving a serious illness or a potentially fatal accident; supernatural causes and cures for illness; different levels of the universe
with respective spirit rulers; man-animal transformations; animal
spirit helpers, alter egos, and guardians; and acquisition of supernatural or "medicine power" from an outside source (Furst 1976: 6).
Also, Hultkrantz (1985) proposed that shamanism is usually found
in loosely organized hunter-gatherer societies and disappears
with a change to more structured horticulture or agriculture. Yet
all the shamanistic characteristics on Furst's list were present in
Aztec society coincident with state-level institutions and an ecclesiastical religion. Schultes and Hofmann (1979: 27–30) pointed out
that Mexico represents the world's richest diversity in the use of
hallucinogens by aboriginal societies despite a comparatively
modest number of species in the flora. This is significant in that
hallucinogens are the most common means of inducing an altered state of consciousness such as the ecstatic trance that is a
key component of shamanism. La Barre (1970) proposed that the
original migrants who settled the New World were hunter-
gatherer practitioners of shamanism and therefore culturally pro-
grammed to seek hallucinogens in their new environment and to
adopt local plants for such use. In contrast to the rest of the world,
New World religions retained their shamanistic traits even societies that, like the Aztecs', achieved a state level of complexity.

Among the Aztecs, shamans (*nanahualtin—nahualli*, singular)
were predestined. People born on a day 1-rain or 1-wind would
become nanahualtin (Sahagún 1950–1969: book 4, 93, 101).
Nanahualtin were chosen by illness vocations, by being narrowly
missed by a lightning bolt, or by being marked at birth by albinism,
lameness, cross-eyes, or other birth defects (de la Serna 1953: 86–
87, 204–242). It was said that a nahualli at birth would repeatedly
emerge from and retreat into the womb, leaving the womb four
(*naui*) times. We have already discussed the three levels of the universe, with their different rulers, and the fact that tonalli could
leave the body. These levels also corresponded to different time
scales: the first was the time of the gods; the second was the time of
creation, mythic time; and the third was that of this world, the

earth time of humans. Shamans differed from ordinary people in that they could voluntarily send their tonalli on magical flights to other worlds outside their body, aided by hallucinogens. A shaman could travel to mythic time, in either the upper or underworld, and simultaneously see his tonalli operate in earth time (López Austin 1988: vol. 1, 61–62, 67). The tonalli could enter animals (animal doubles of the shaman) also called nahualli. A god could also appear as a nahualli, a sort of disguise. For example, Tezcatlipoca's nanahualtin included the skunk and the coyote, while Painal's, an avatar of Huitzilopochtli, was a hummingbird (López Austin 1988: vol. 1, 368).

Texts on drunkenness mention extracts from the cacti or herbs *peyotl* (*Lophopora* sp.), *tlapatl* (*Datura* sp.), and *ololiuhqui* (*Rivea* sp.) and use terms such as *itech quinehua* ("it takes possession of him") and *itech quiza* ("it comes out in him"). These indicate the Aztec belief that possession by deities contained in these plants caused their effects. This concept of god "in a plant" is expressed by the word *entheogen* ("God within us") coined by Ruck (Wasson 1980: xiv) to replace terms such as hallucinogens, psychotomimetics, and psychedelics. These plants had a double function. They housed the god that took possession of the consumer and also sent the consumer's tonalli on a magical flight to mythic time and space (López Austin 1988: vol. 1, 356–358). We now know that visions seen under the influence of entheogens are artificial, due to the actions of neurotransmitters such as serotonin. Aboriginal shamans perceived the entheogenic experience as another dimension of reality. Reichel Dolmatoff (1978: 151–152) reported that the Tukano Indians today distinguish between what one knows by seeing in ordinary reality and another type of knowledge that the mind "hears" under the influence of entheogens. Thus one cannot truly comprehend "reality" unless one sees both the real world and the mythic world as perceived in an altered state of consciousness. Wasson (1980: 79–92) proposed that the word xochitl ("flower") in Nahuatl poetry often referred to entheogens, for example, in the poem:

> Our priests, I ask of you:
> From whence come the flowers that enrapture man?
> The songs that intoxicate, the lovely songs?
> Only from His home do they come, from the innermost part of heaven,
> Only from there comes the myriad of flowers.
> (León Portilla 1963: 77)

The word in the poem describing "flowers" as "enrapturing" or "intoxicating" is *ihuinti,* which was also used to describe the actions of hallucinogens. Caceres (1984: 208) commented that entheogens also produce auditory effects that can take the form of music. Therefore, the second word of the Aztec two-word metaphor for poetry, cuicatl ("song"), may also refer to the effects of entheogens. In 1961, when León Portilla (1961: 128) wrote about the Aztec belief that only poetry—in xochitl in cuicatl—could lead to truth, and that truth "was not found on earth," much less was known about the entheogenic experience than is known today. It is plausible that the metaphor was based on shamanic experience and refers, as in the case of the Tukano, to the idea that the full meaning of reality can be perceived only under the influence of entheogens.

Shamans were the agents of choice for curing the kinds of ailments caused by disequilibrium of the universe or the intrusion of beings from other cosmic levels. Diagnosis and cure of these ailments required that the healer travel to these other worlds. López Austin explained this process (1984: 113):

> The voyage to the cosmos was accomplished through ritual and the ingestion of psychotropics, which resulted in the physician being called *pahini,* that is, "he who consumes medicine." In some rituals, the magician-doctor set up his place on the surface of the earth, reproduced physically by a mat on which a fire symbolized the center of the earth and four objects placed at the corners represented the cosmic pillars by which the gods and their influences ascended and descended. His access to the forbidden places took place by the same routes used by the gods, and the physician called himself—"I the traveler who goes to the World of the Dead, I the traveler who goes to that which is above us" *(niani Mictlan, niani topan).* He needed to have knowledge of the *nahuallatolli,* the particular language used to communicate with the invisible beings. By having his *tonalli* leave his body and locate itself in either the upper or the lower world, the physician was able to confront time freed from the constraints of the reality of the intermediate sector of the cosmos [that is, the earth]. There, he could search and discover the cause of the illness, perceive the prognosis, and operate in mythic time to affect the supernatural factors which were responsible for the pathology of his patient. He then returned to the middle sector of the cosmos with what he had acquired, and there he was able to act with his new knowledge or deal with a condition which he had modified by his actions in the other world.

Shamanic curing and the use of entheogens are still widespread among native groups in Mesoamerica. These groups, as did the Aztecs, believe in predestination and illness vocation. They also believe that the power to cure is obtained by trips to other worlds in altered states of consciousness.

The Aztec religion and worldview permeated all aspects of Aztec society. They were a key component of their approach to illness: its causes, its cures, and its prevention and amelioration. Thus, Aztec society was unusual in several respects; the threat of illness was used as a powerful instrument of social control to promote moderation and performance of duty; strong shamanic elements were present despite an official state-level religion; and a reciprocal correspondence and influence existed between man and the cosmos.

3

Population and Carrying Capacity of the Basin of Mexico

Nutritional status is a very important variable in evaluating the health of a population. An adequate and well-balanced diet supports the body's immune system as the first and most important defense against infectious and parasitic diseases. Evolutionarily and historically the immune system has been the primary determinant of health in population. Until the widespread use of antibiotics in the 1940s, Western medicine, like other medical systems, could only administer palliative measures and try to bolster the immune system when dealing with infectious diseases. Even today, failure of the immune system (as in the case of AIDS) is a death sentence despite the advances of biomedicine. Therefore, it is important to assess the nutritional status of the Aztecs. This chapter reviews arguments concerning the population density and carrying capacity of the Basin of Mexico and the recent proposal that Aztec cannibalism was caused by an inability to obtain sufficient protein from other sources, and concludes that the population did not exceed the Basin's carrying capacity. Chapter 4 discusses the nutritional adequacy of the Aztec diet and the Aztecs' use of certain little-known plants that contributed greatly to both agricultural productivity and dietary nutrition.

Historians strongly disagree about the nutritional adequacy

of the Aztec diet. This controversy, and closely linked questions about Aztec population figures, can be considered in terms of the differing European views of the Aztecs, described in chapter 1, which date from the Conquest. Western opinions have continued to oscillate between a high and a low estimate of Aztec accomplishments.

In the area of food and nutrition, two viewpoints prevail today. One holds that the Aztecs before the Conquest were undernourished, malnourished, or both, and that the arrival of the Europeans resulted in an improvement of the native diet. Cook and Borah (1979; Borah 1980) are proponents of this view. Harner (1977) and Harris (1978) went to the extreme of proposing that cannibalism was an Aztec solution to a desperate need for protein. An opposite view, to which I subscribe, is that the Aztecs had access to excellent food resources and good intensive agricultural techniques and thus managed to feed the population a well-balanced diet better than that of many poor Mexicans today. Some other proponents of this second view are Sanders (1976b; Sanders et al. 1970), Santley (Santley and Rose 1979), and Ortiz de Montellano (1978).

Conclusions about the adequacy of the native diet depend to a large extent on the estimated population of Central Mexico on the eve of the Conquest. The larger the population, the more difficult for it to have received adequate nutrition. On the basis of their estimates of the number of inhabitants (relatively high) and agricultural production in Central Mexico in 1519, Borah and Cook concluded that, although the diet eaten by the Aztecs was qualitatively adequate in protein and vitamins, quantitatively commoners existed in a state of compensated undernutrition, that is, inadequately fed but able to function (Borah and Cook 1963; Cook and Borah 1979: 163). Harner used the same high population estimate to underpin his argument for the necessity of cannibalism. We shall briefly discuss these population estimates, including Sanders's opposing figures, and compare them with the proposed regional carrying capacity, focusing on the area surrounding Tenochtitlan because it was the most densely populated and would have posed the severest test of the Aztec ability to feed a large population.

Comparisons between the Sanders and the Borah and Cook estimates are complicated because they do not apply to strictly comparable land areas. For example, Borah and Cook defined "Central Mexico" as the area between the Isthmus of Tehuantepec and the northern border of sedentary [Indian] settlements in 1550 (Borah and Cook 1963: 3), about 514,000 km² inhabited by about 25,200,000 people, making a population density of 49/km². Their Region I subdivision included the entire modern states of Mexico, Tlaxcala, Morelos, Hidalgo, most of Puebla, the north quarter of Guerrero, a small part of Veracruz, and Mexico City, an area of 202,439 km² (Borah and Cook 1963: 157) with a total population of 10,900,000, including Tenochtitlan, for a density of about 54/km² (Borah and Cook 1963: 79). Sanders (1976a: 86–87) defined a Central Mexican Symbiotic Region, including the Basin of Mexico and adjacent areas in southern Hidalgo, southwestern Tlaxcala, the western third of Puebla, and the state of Morelos, with an area of 20,811 km² for which he estimates a population of between 2,600,000 and 3,100,000 and a density of 125–149/km². This core area had intensive agriculture and a heavily concentrated population, including the city of Tenochtitlan. Table 3.1 lists these and other populations adjusted by Sanders (1976a) and grouped to make his and Borah and Cook's estimates more comparable. When ranges were given, only their midpoints are cited.

Sanders (1976a) presented a thorough critique of Borah and Cook's methodology and results and believed they are unrealistic. Particularly telling is his calculation that the Valley of Teotihuacan could have supported between 150,000 and 170,000 people with a diet based on maize; Borah and Cook estimated that 320,000 people lived there in 1519 (Sanders 1976a: 136–142). We will do a similar calculation for the Basin of Mexico, because it was the area of greatest population density and most intensive agriculture, in order to evaluate the two conflicting population estimates.

The borders of the Basin of Mexico are defined slightly differently in different sources. For consistency, we shall use an area of 7853 km² (Kovar 1970: 15). At the time of the Conquest, a series of shallow lakes within the Basin covered an area of approximately 1,000 km², most of it occupied by salty Lake Texcoco. Kovar (1970: 15) divided the Basin area into different zones based on altitude: lake (1,000 km²), below 2,250 m (507 km²), hilly zone below 2,400 m (2,675 km²), intermediate zone below 2,700 m (2,671 km²), and mountain zone above 2,700 m (1,000 km²). This division has significant ecological implications because despite the region's tropical location, it does not have a tropical climate due to its high altitude. The Basin of Mexico is characterized by a rainy season of five months beginning in May and ending in September, which accounts for seventy-five percent of the total annual rainfall. Pre-

Table 3.1. Estimated Populations of Various Areas of Central Mexico, 1519

| | CENTRAL MEXICO | | CENTRAL MEXICO | SYMBIOTIC REGION | MEXICO BASIN | |
	Total	Per/km²	Total	Per/km²	Total	Per/km²
Borah & Cook	2.52×10^7	49	$6.4 \ \times 10^6$	308	2.56×10^6	330
Sanders	1.14×10^7	22	2.85×10^6	136	1.16×10^6	149

Source: Sanders, 1976a: 88, 120, 130.

cipitation varies from an annual average of 450 mm on the northeastern plains to more than 1,500 mm in the southwestern mountains (Sanders, Parsons, and Santley 1979: 82). A serious problem the Aztecs faced is corn's low resistance to frost in the Basin because the plant species had evolved at lower, and warmer, altitudes. It is particularly vulnerable at the critical stages of germination and ear formation when, furthermore, its moisture requirements are high:

> The major problem, therefore, is the relative timing of rain and frosts: A late rainy season and early frosts are fatal for the maize crop. The frost-rainfall problem varies considerably in significance from area to area within the Basin. Within the elevation band where maize can be grown (2240–2700 m), the frost-rainfall problem is most severe in the upper piedmont (2500–2700 m) and the outer alluvial plain (2250–2300 m)—two areas which together make up 40% of the surface area. The problem is less severe in the south, where the precipitation both is more abundant and more regular. Of the remaining 45% of the Basin, 29% lies below 2250 m (two-thirds of this was lake, and the rest is a narrow band along the lakeshore); the other 25% lies between 2300 and 2500 m, the lower middle piedmont. Both the immediate lakeshore plain and the lower middle piedmont are relatively free of the frost problem. (Sanders, Parsons, and Santley 1979: 222)

The total amount of rainfall can also be a problem in certain parts of the Basin. Theoretically, because most of the rain falls during the warm months of the growing season, conditions should be favorable for agriculture. In fact, the rainfall is very variable from year to year and frequently is inadequate for a good crop. For example, during the last sixty years, rainfall dropped below 500 mm thirteen times at one measuring station (Sanders, Parsons, and Santley 1979: 225). The major solution to the frost-rainfall problem is irrigation and planting before the beginning of the rainy season. The ideal arrangement is to irrigate in March or April and to plant in April or early May. After planting, crop growth can depend entirely on natural rainfall and avoid the danger of early fall frosts. In this manner, a small amount of water can be used very effectively to ensure stable crop production. Irrigation systems can also be used to supplement natural rainfall if necessary.

Harner (1977) claimed that frequent droughts and famines led the Aztecs to cannibalism to supplement their diet, and others

(Hernández Rodríguez 1962; Bustamante 1975; Viesca Treviño 1982), cited in apparent support of this contention, instances of drought and famine in various codices and chronicles. However, these reports must be placed in context to truly evaluate their significance. For one thing, most Mesoamerican historical codices deal with events in small locales rather than with area-wide histories. Also, as Sanders, Parsons, and Santley (1979: 231) wrote:

> rainfall is highly localized in the Basin so that one area may have a highly favorable rainfall-temperature regime in a particular year while another suffers a disaster, even though the two are only 15–20 km apart. As a result, it would be unlikely that drought phases would be generalized in timing and duration over the entire area served by a large irrigation system.

Cox (1978) cited a pattern of much more frequent famines recorded in the Old World than in the New World and concludes that the contrast reflects real ecological differences and not just differences in record keeping. He pointed out, as we will emphasize below, that peoples in the New World not only practiced agriculture but also retained highly diversified food procurement systems, including hunting and gathering. Hassig (1981: 178) agreed that famines were relatively rare in the New World, particularly in contrast to the famine belts of the Old World.

One exception to the New World's relatively favorable experience stands out: 1-rabbit (1450–1455 A.D.) was the most devastating Aztec famine recorded and clearly led to disease, massive migration, and widespread death. Beginning in 1450, the Valley of Mexico suffered crop-destroying frosts, followed by two years of severe drought, but the greatest suffering and the abandonment of the city of Tenochtitlan did not occur until 1454–1455 (Hassig 1981). Sources mention that the ruler Motecuhzoma staved off disaster for the first two years by distributing stored food previously obtained in tribute. Only when this resource was exhausted did the drought truly manifest its worst consequences. Hassig (1981: 175) took this as evidence that the population of the Valley of Mexico did not normally strain the Valley's carrying capacity and that the sustained production failure of 1-rabbit resulted from an unusual circumstance of several years of climatological misfortunes.

The catastrophe provoked a series of reactions from the Aztecs in an attempt to prevent its recurrence. After 1455, a great expansion of public works was initiated to promote intensive, irrigated, and droughtproof agriculture. The main focus was on *chinampas,* artificial planting areas built from mud scooped from the bottom of the lake, held in place by posts and roots, and separated by canals carrying water used to irrigate the chinampas. Chinampas are extraordinarily productive. Continuously cropped, they produce up to seven crops per year by the extensive use of seed beds and crop rotation. They are obviously droughtproof, standing in the middle of a lake. One person can work 0.75 hectares (ha) of chinampas and produce 3,000 kg of corn per year (Parsons 1976). The estimated area of chinampas at the time of the Conquest was 10,000 ha in the fresh-water Lake Chalco and 2,500 ha in reclaimed areas of salty Lake Zumpango. This great expansion of chinampas in response to the famine of 1-rabbit was made possible by a dike 16,000 m long built to allow effective control of the level and salinity of the southern lakes.

In other parts of the Basin, the crisis was resolved by conversion from swidden (slash-and-burn) agriculture to intensive irrigation and terracing. The conversion involved the construction of both canals for permanent irrigation and dams to allow irrigation with floodwater from rain (Palerm and Wolf 1972: 121–126). Sanders, Parsons, and Santley (1979: 380) estimated that 59 percent of cultivable land was irrigated in some form.

= Carrying Capacity of the Basin of Mexico and Food Consumption

The plausibility of Borah and Cook's population estimate in table 3.1 can be tested by determining the carrying capacity of the Basin of Mexico and comparing it with the proposed population. A population that greatly exceeds the carrying capacity could not be sustained and is improbable.

The biological definition of carrying capacity is the maximum number of individuals of a species that an area can sustain without long-range deleterious effects on the area's capacity to continue doing so. This definition poses a problem when applied to humans who, through culture, can change their food production techniques to sustain larger populations in the same area. However, in the 1500s, the Basin of Mexico was already being exploited intensively, so a calculation of carrying capacity for the agricultural techniques of that time can serve the limited purpose of evaluating the reasonableness of various population estimates and adequacy of nutrition. Sanders, Parsons, and Santley (1979:

372–385) used three factors in calculating carrying capacity: the cultivation factor, the land-use factor, and the cultivable-land factor. The cultivation factor refers to the planted land area necessary to sustain an average person in a particular year. It is composed of two subsidiary factors: land productivity (kg/ha) and food consumption (kg/person). Sanders, Parsons, and Santley combined both subsidiary factors into one number and changed the base from one person to an extended family of 7 people. Borah and Cook used an average family size of 4.5 as their base. Here, we will keep the two components separate and calculate the cultivation factor on the basis of one person to facilitate comparisons between the different authors.

Maize is the crop traditionally used for productivity and consumption calculations because it was and continues to be the major regional food staple, and because its production has been well studied. Sanders, Parsons, and Santley (1979: 373) presented detailed examples of yields under various soil and humidity conditions, but we shall use only a few of these. So many assumptions are involved in determining the amount and characteristics of land available for cultivation that subdivision into numerous types adds confusion without much improving what is an admittedly approximate final estimate. Assumed corn yields are as follows: irrigated land—1,400 kg/ha, temporal alluvial plain—1,000 kg/ha, temporal piedmont—600 kg/ha, fallow bush—1,400 kg/ha, and chinampa—3,000 kg/ha. These are conservative values, considering that Patrick (1977: 201) showed an average production of 4,000 kg/ha on unirrigated terraced lands in this region.

To merit consideration, a calculation of the amount of food consumed per person must be founded on clearly stated assumptions. Borah and Cook, and Sanders, based their calculations on modern caloric requirements modified by assumptions about family composition and/or metabolic requirements. On the basis of historical data, Borah and Cook (1963: 90) also proposed that calories were totally derived from 958 g of maize (3,350 cal) per day per person, or 350 kg/yr. Adjusted for the differing requirements of men, women, and children in their average family, this is equivalent to 256 kg/yr of maize per person. In a more recent calculation (Cook and Borah 1979: 140–164), taking into account body size, size-related basal metabolism, and added demands of various degrees of physical labor, as well as the differing requirements of men, women, and children, the estimate was lowered to 543 g (1,900 cal) per day per person, or 198 kg/yr. Borah and Cook

Table 3.2. Estimates of Aztec Corn Needs per Person in 1519

AUTHOR	TOTAL CAL/DAY	% FROM CORN	KG CORN/YR
Borah & Cook 1963	3,350	100	256
Cook & Borah 1979	1,900	100	198
Cook & Borah 1979	1,400–1,800	100	146–188
Sanders 1976a	2,000	80	160
Ortiz de Montellano 1978	2,629	69	186

claimed that the natives existed in a state of compensated mal-nutrition, actually consuming a diet of only 400–515 g/day (1,400–1,800 cal), or 146–188 kg/yr, still derived totally from maize.

Sanders (1976b: 109) suggested a diet of 2,000 cal/day per person, 80 percent of which, by A.D. 1500, was being supplied as maize, requiring 160 kg/yr per person. Ortiz de Montellano (1978) estimated consumption at 2,629 cal/day (734 g if derived from maize), significantly more than the 2,200 cal/day recommended by the World Health Organization, on the basis of a balanced diet composed of foodstuffs given to Tenochtitlan in tribute augmented by maize as the limiting factor. On this diet, an average person would derive 69 percent of total calories by eating 186 kg of maize (and amaranth—see chap. 4) per year. These calculations are summarized in table 3.2. Cultivation factors can be calculated for the various consumption estimates by dividing the annual consumption per person (kg of maize) by the average annual productivity (kg/ha) of the cultivated land areas. However, carrying capacity requires consideration of the other contributing factors (see table 3.3).

The second component in the calculation of carrying capacity, the land-use factor, is the total land area needed to continuously support a family or individual, expressed as a multiple of a unit area equal to the cultivation factor. It includes the land under current cultivation, needed to supply the required diet in any one year, plus the number of unit areas of land left fallow to ensure meeting future needs for continuous sustenance. Irrigated land and chinampa can be used every year (land-use factor = 1); temporal, alluvial plain, and piedmont can be used every other year

Table 3.3. *Annual Corn Yield, Basin of Mexico, 1519*

CULTIVABLE LAND		AREA, HA	LAND-USE FACTOR	AREA IN CURRENT USE, HA	ANNUAL YIELD KG/HA	ANNUAL YIELD, KG
Low (below 2,250 m)		5.07×10^4				
Irrigated (50%)		2.54×10^4	1	2.54×10^4	1.4×10^3	3.56×10^7
Not irrigated		2.54×10^4	2	1.27×10^4	1.4×10^3	1.78×10^7
Hilly (below 2,400 m)		1.77×10^5				
Irrigated (25%)		4.43×10^4	1	4.43×10^4	0.9×10^3	3.99×10^7
Not irrigated		1.33×10^5	5	2.66×10^4	1.4×10^3	3.72×10^7
Intermediate (below 2,700 m)		1.34×10^5				
Irrigated (25%)		3.35×10^4	1	3.35×10^4	0.9×10^3	3.02×10^7
Not irrigated		1.01×10^5	5	2.02×10^4	1.4×10^3	2.83×10^7
Chinampa		1.25×10^4	1	1.25×10^4	3.0×10^3	3.75×10^7
Totals	(cultivable)	3.74×10^5 ($3.74 \times 10^3 \text{ km}^2$)		1.75×10^5		2.27×10^8

Notes: Land division by elevation according to Kovar (1970). Cultivable areas exclude a portion of the total; low land omits 10^5 ha of lake surface, hilly land is assumed to be 66 percent cultivable, and intermediate land is assumed to be 50 percent cultivable.

Table 3.4. Carrying Capacity Estimates, Basin of Mexico, 1519

AUTHOR	ESTIMATED CARRYING CAPACITY	NO. OF PEOPLE
Borah & Cook 1963	886,718	(8.6×10^5)
Cook & Borah 1979	1,146,464	(1.1×10^6)
Cook & Borah 1979	1,554,794–1,207,446	$(1.6–1.2 \times 10^6)$
Sanders 1976a	1,418,750	(1.4×10^6)
Ortiz de Montellano 1978	1,220,430	(1.2×10^6)

(factor = 2), and fallow bush can be used every five years (factor = 5) (Sanders, Parsons, and Santley 1979: 376).

The third component, the cultivable-land factor, is the arable percentage of the total land area. In the Mexican case, the key consideration is altitude, which determines the occurrence of fall frosts. Sanders, Parsons, and Santley (1979: 378) estimated that 80 percent of the alluvial plain, two-thirds of the lower piedmont, and half the middle piedmont could be cultivated: land higher than 2,700 m was not cultivable. These figures may underestimate the arable area. Patrick (1977: 246–250) suggested that we may be biased in our view of what constitutes agricultural land. For example, we favor flat surfaces over terraces, which the Aztecs used extensively. Table 3.3 summarizes the calculation of the annual corn yield for the Basin of Mexico in A.D. 1519. Land is grouped by altitude as defined by Kovar (1970). Productivity estimates and cultivable-land factors are given in the table and accompanying notes.

Table 3.4 lists various estimates of the carrying capacity of the Basin of Mexico in 1519 based on corn yields shown in table 3.3 and dietary needs proposed by cited authors.

A comparison of tables 3.1 and 3.4 shows that the Borah and Cook population estimate of more than 2,500,000 for the Basin of Mexico in table 3.1 is clearly unreasonable. It exceeds by about a factor of two even their own revised figures (1979), based on a diet that led to compensated malnutrition. Sanders's estimate of 1,160,000 in table 3.1 does not exceed our approximate minimal carrying capacity (table 3.4) and thus warrants confidence. A recent unpublished critical review of the literature, which attempted a population reconstruction for the Basin of Mexico (Whitmore and Turner n.d.), concluded that 1,200,000 was the most reasonable estimate for 1519, a figure we can comfortably assume.

The response of the Aztec Empire to the famine of 1450 involved intensification not only of agriculture but also of military activity in order to conquer new areas for potential tribute. The direction of expansion was also changed, tending more toward areas with greater assumed rainfall and high fertility, areas where Aztec children had been sold for food during the great famine (Ortiz de Montellano 1978). Hassig (1981; 1985: 129) objected to accepting this response as a plausible solution to the problem of feeding the population of the Basin of Mexico, on the basis that transport required human carriers who, in turn, consumed food. He maintained that corn and bulk foodstuffs could be efficiently imported only from territories one day's travel away. According to his theory, the additional area to be included in calculations of the carrying capacity of the Basin of Mexico, if food tribute were taken into account, must lie within one day's travel of the *tlameme*, the professional porter. Hassig (1985: 40) calculated that a tlameme could carry a load of 23 kg a distance of 21–28 km/day.

Hassig's proposal raises several problems. One is that bulk food tribute was in fact carried longer distances than he proposed as a limit. For example, 500 tons (5,000 kg) of corn, beans, and amaranth were brought yearly to Tenochtitlan from the province of Tepecoacuilco, a distance of 240 km, which would have required a 16-day round trip (Litvak King 1971: 111–112). Food tribute also came some 340 km from the remote province of Coyolapan (Berdan and Durand-Forest 1980: 28, 39). Another problem is that Hassig's estimated cost of transport seems excessive. He maintained that tlamemes, because they were professionals, were paid enough corn to sustain their entire families, assumed to average five people, and included in the cost of a one-way delivery the food required for the return trip. He sets the cost of transport at 30 percent of the load itself (Hassig 1985: 129); at 8 lb (3.63 kg) of corn a day for a tlameme's family of five, this amounts to a payment of 7.26 kg of corn for a round trip to deliver a 23-kg load. It also assigns a very high caloric intake to the family, at 4,000 cal/kg, 3.63 kg of corn provides 14,520 cal, or 2,904 cal/day for each member of a family of five.

There is no reason why a tlameme would have to return empty-handed. He would more logically return to the provinces carrying articles for trade, so the entire cost of the trip should not be charged to transport of the food cargo. Also the food requirements proposed by Hassig can justifiably be lowered. An adult male with a normal basal metabolism would require 2,365 cal/day to do agricultural labor (Cook and Borah 1979: 157). Adding calories to accommodate the harder work of the tlameme yields a higher daily requirement for the tlaleme, but the total requirement for a family of five is reduced from Hassig's 3.63 kg to 2.26 kg of corn/day, a drop of over 35 percent, if the dietary needs of the other members are also derived from basal metabolism.[1] Using this figure and charging only a one-way trip to the delivery of food reduce the cost of transport to 9.8 percent of the load per day and increase the radius of the effective support area to 63–84 km.

The Basin of Mexico is also unique in that food could effectively be delivered to Tenochtitlan at any point on the lakes' shoreline and the journey completed by canoe. Since a canoe's load-carrying capacity has a 40:1 advantage over that of a tlameme, the cost of the final leg of the journey is minimal (Hassig 1985: 64, 66). We can calculate the effective economic area of Tenochtitlan as extending either 5 leagues (28 km) from the shoreline, as Hassig proposes, or 15 leagues (84 km), if we use the revised calculations above.[2] At 5 leagues, the effective economic area is 5,600 km^2, or 83 percent of the area of the Basin of Mexico. At 15 leagues, it is 31,600 km^2, which is larger than the Central Mexican Symbiotic Region outlined by Sanders (1976a: 131). What this means, even at the lower figure derived from Hassig's assumptions, is that all the land area of the Basin and its food production can be used in calculations of carrying capacity without the specter of a transportation bottleneck becoming a limiting factor.

Such calculations of carrying capacity indicate the order of magnitude of the possible population but are not precise because they assume an all-maize diet. The actual diet of the Aztecs was greatly enhanced, both quantitatively and qualitatively, by a number of foods not usually included in calculations, some of which have even been discounted as possible dietary supplements. One of the foods usually excluded is the alga *tecuitlatl* (*Spirulina geitlerii*), which Santley and Rose (1979) suggested made possible the growth of population in the Aztec period. Examples of discounted foods are wild animals and insects which, contrary to the opinion of Harris (1979a: 336–340), made significant qualitative contributions to the diet (McGregor 1984; Sahlins 1978).

⸗ Aztec Cannibalism

Harner (1977) contended and Harris (1978) concurred, that Aztec cannibalism was a response to a need for protein. This author (Ortiz de Montellano 1978) and others (Price 1978; Garn 1979) have denied the validity of this contention. A list of Harner's arguments follows, accompanied by brief rebuttals.

Argument: The Aztecs lacked domesticable hervibores as a good source of protein.
Rebuttal: As we shall see, the Aztecs had available an extraordinary range of animal protein and had domesticated and consumed turkeys and dogs. Hunn (1982) lists "domesticable" herbivores (peccary, pronghorn antelope, tapir, agouti, and muscovy duck) in Mesoamerica and claims that Aztecs could have domesticated them but deliberately chose a more efficient means of providing nutrients through intensive agriculture.

Argument: A corn diet does not contain all the essential amino acids. A combination of corn and beans does, provided they are eaten together, but this was not necessarily done.
Rebuttal: It is ethnocentric to believe that only meat consumption constitutes a satisfactory diet. As we shall see, customary Aztec diets satisfied all nutritional requirements, including vitamins and minerals, and the quality of their protein was quite satisfactory.

Argument: The Basin of Mexico was subject to frequent droughts that caused food shortages, famines, and consequent increased population pressure. Increased population pressure led to more human sacrifice accompanied by cannibalism to remedy the protein shortage.
Rebuttal: As shown above, the regional rainfall-temperature pattern tended to make droughts local rather than widespread. Famine in Mexico was less frequent than in the famine belt of the Old World (which did not turn to cannibalism when hungry). The really serious, and unusual, famine of 1450–1455 provoked a series of responses that dealt effectively with the problem.

Argument: Harner cited Cook and Borah's estimate of a population of 25 million and Borah's unpublished figure of 250,000 yearly

sacrifices for Central Mexico and corresponding figures for Tenochtitlan of 300,000 people and 15,000 yearly sacrifices. In Tenochtitlan, only the limbs of the victims were consumed, and only
the upper class (approximately 25 percent of the population) was
allowed to partake. The rest of the population are claimed to have
supported the practice of warfare and sacrifice because bravery in
combat was a possible means of becoming part of the privileged
nobility and sharing their access to the extra food.
Rebuttal: Cook and Borah's estimate for the population of Central
Mexico is not reasonable and should be cut in half; therefore, the
population pressure was much less than Harner envisioned. The
estimated number of human sacrifices should also be halved,
since it was an assumed fraction of the proposed population
rather than proven independently.

We shall show that the diet of the Aztecs was nutritionally adequate not only in calories and protein but in all other requirements. If protein was not lacking, the crucial part of Harner's
argument is obviously invalidated. We shall also contend that if
protein actually was vitally needed, cannibalism, even at the scale
postulated, would not have been a sufficient remedy. Finally, we
shall maintain that cannibalism reached its peak not at a time of
potential food scarcity in the agricultural cycle, when it might
conceivably have alleviated hunger, but at harvest time (Ortiz de
Montellano 1978). Most of the data presented will deal with Tenochtitlan, both because more information is available for this city
and because Harner cited it as the locale of the most extreme use
of sacrifice (five percent of the population yearly compared to one
percent for Central Mexico).

Large amounts of corn, beans, chia (*Salvia hispanica*), and
amaranth (*Amaranthus sp., huauhtli*) were brought yearly as tribute to Tenochtitlan. These four staple foods are an adequate basis
for diets that meet or exceed the recommended daily allowance
for calories, proteins, and other nutrients. Two such diets, shown
in table 3.5, which include the grain amaranth in addition to corn,
exceed the Food and Agricultural Organization-World Health Organization (FAO-WHO) requirements for protein and calories, as
well as the caloric requirements proposed by Sanders and Ortiz
de Montellano (table 3.2) but not those of Borah and Cook (1963).
Harner's overemphasis on protein, particularly *meat* protein, is
misplaced. The crucial factor in preventing malnutrition is an adequate total supply of calories of which only at least five percent

Table 3.5. *Sample Daily Diets from Aztec Tribute Grains*

	ENERGY kcal	PROTEIN g	Ca mg	P mg	VIT. A mg	THIAMINE mg	RIBOFLAVIN mg	NIACIN mg	VIT. C mg
Diet 1									
Corn (400 g)	1,432	33.6	44	484	0.6	1.52	0.4	7.6	—
Beans (100 g)	343	22.7	1.3	415	0.008	0.47	0.4	2.1	1
Chia (100 g)	463	15.6	518	518	0.01	0.38	0.13	3.74	—
Huauhtli (100 g)	391	15.3	490	455	—	0.14	0.32	1.0	4.5
Total	2,629	87.2	1,053	1,872	0.62	2.51	1.25	14.4	5.5
Diet 2									
Corn (300 g)	1,074	25.2	33	363	0.45	1.54	0.3	5.7	—
Beans (200 g)	686	45.4	2.6	830	0.02	0.94	0.3	4.2	2
Chia (200 g)	926	31.2	1,036	1,036	0.02	0.76	0.26	7.48	—
Huauhtli (100 g)	391	15.3	490	455	—	0.14	0.32	1.0	4.5
TOTAL	3,077	117	1,562	2,684	0.49	3.38	1.18	18.4	6.5
FAO-WHO	2,200	45	800	800	1	1.2	1.8	20	45

Source: Data from Flores et al. 1960; Becker et al. 1981. Ortiz de Montellano 1978, copyright © 1978 by the AAAS.

need be derived from protein (Waterlow and Payne 1975). Most cereals provide usable protein at close to this level. In self-selected diets in most areas of the world the quantity not the quality of the food supply is the limiting factor in the supply of protein (Beaton and Swiss 1974; Payne 1975). We must also remember that the Aztec grain diets were usually supplemented by various vegetables, fruits, insects, and wild game so that the total intake was greater than shown in table 3.5. The question of protein quality and limiting amino acids will be discussed later. We should point out, for now, that the amino acid composition of amaranth compensates for the small amount of lysine in corn.

Table 3.6 lists the annual amounts of tribute to Tenochtitlan of the major four grains and the number of people who could be fed for one year at the diet levels shown in table 3.5. Thus, solely on the basis of the four primary grains levied as tribute, between 40,000 and 150,000 people could be fed a balanced diet exceeding FAO-WHO requirements.

Authors generally agree that cannibalism has never served as the principal source of protein for human diets (Garn and Block 1970; Walens and Wagner 1971; Vayda 1970; Dornstreich and Morren 1974). Opinions differ about the importance of smaller contributions. Garn and Block (1970) judged that consumption of less than one adult human body per week (fifty-two per year) per group of sixty persons (equivalent to eating a number of humans each year equal to eighty-seven percent of the consuming population) would not be significantly beneficial even as a supplement to a cereal or tuber diet. Dornstreich and Morren (1974), on whom Harner relied, argued that a contribution from cannibalism ranging from five to ten percent of protein need would be of significant benefit, comparable to the contribution made by pork in New Guinea. To reach this level a population of one hundred (forty-six adults and fifty-four children) would have to eat ten to fifteen adults per year. It should be pointed out that in New Guinea, the chief source of calories is tubers with a very low protein content, under the necessary five percent of total calories. This is not comparable to the circumstances in Mesoamerica, where the staple foods are cereals that supply more than five percent as protein energy. Nevertheless, we shall use these percentages to judge the effectiveness of Aztec cannibalism as a dietary supplement.

Vayda (1970) argued that cannibalism may be a critical source of protein to individuals who are wounded or under severe stress, conditions that deplete body protein. Harris (1979b) also said that

Table 3.6. *Population Dietically Supportable by Tribute*

Conversion factor (kg/fanega)[a]	42	55.5	75	119
Corn				
Weight (10³ ton)[b]	5.90	7.77	10.5	16.7
Diet 1 (10³ persons)[c]	40.4	53.2	71.9	114
Diet 2 (10³ persons)	53.9	71.0	95.9	152
Beans				
Weight (10³ ton)	4.41	5.83	7.88	12.5
Diet 1 (10³ persons)	121	160	216	343
Diet 2 (10³ persons)	60.4	79.8	108	171
Chia				
Weight (10³ ton)	4.41	5.83	7.88	12.5
Diet 1 (10³ persons)	121	160	216	342
Diet 2 (10³ persons)	60.4	79.8	108	171
Huauhtli				
Weight (10³ ton)	3.78	5.00	6.75	10.7
Diets 1, 2 (10³ persons)	104	137	185	293

Source: Ortiz de Montellano 1978, copyright © 1978 by the AAAS.

[a]Weight depends on the estimated number of kilograms in a fanega, a Spanish dry measure now equal in Spain to 1.58 U.S. bushels.

[b]Weights are given in metric tons = 1,000 kg.

[c]Number of persons (in thousands) who could be fed for one year at the daily level for each diet given in Table 3.5.

trauma or debilitation by infections or wounds increases the body's protein requirement and suggests that cannibalism might have given elite Aztecs "extra" protein in times of such need. He claimed that this, rather than general dietary improvement, was the real value of consuming human flesh. This is too simplistic a view of the metabolic process. As Waterlow, Golden, and Perou (1977) pointed out, postoperative nitrogen loss (analogous to loss due to a wound) is associated with a considerable decrease in protein synthesis and little change in the rate of protein breakdown. Thus, since the injured body is manufacturing less protein, increased protein in the diet will be wasted because it will not be used.[3]

The amount of protein available to Aztec nobles from cannibalism can be calculated and depends on which of two assumptions is made. Either the entire body was eaten or, as Harner claimed, only the extremities. The latter assumption is probably correct. For the sake of argument and a conservative estimate, we will use the population figures cited by Harner, although we have shown them to be incorrect. Other neglected errors will also inflate the estimates. For example, many women and children were sacrificed, but only a heavier adult male body weight is used in the calculations. Furthermore, victims sacrificed in the many festivals dedicated to the Rain God, Tlaloc, during the year were usually buried intact (Motolinía 1971:63, 66), reducing the available supply of human flesh.

Protein needs of the nobility (25 percent of the total population) allowed to partake of human flesh are calculated by multiplying the daily protein requirement of 0.71 g per kilogram of body weight by an average body weight of 60 kg per consumer and then multiplying the product by the number of consumers and by 365 days per year (Ortiz de Montellano 1978). Tenochtitlan's consumer population of 75,000 nobles would have required approximately 1.2 million kg of protein per year. Protein requirements for Central Mexico, with a consumer population of 6.25 million nobles would have been 97 million kg per year.

Calculation of the protein available from sacrifices depends on several assumptions. We conservatively assume that all victims were 60-kg males, 16 percent protein—similar to 16.5 percent in lean beef and lamb or 14.5 percent in lean pork (Consumer and Food Economics Research Division 1963)—and 90 percent digestible. Skillful butchering would give a 60 percent dressed yield

(Garn and Block 1970). Our assumed victim would therefore yield 60 kg × 0.16 × 0.60 × 0.90 = 5.18 kg of digestible protein. Extremities would equal 35 percent of the total weight, if buttocks and shoulders were included (Spitz 1977). If only the extremities were eaten, the protein yield would be reduced to 0.35 × 5.18 = 1.81 kg. The potential contribution of cannibalism to the diet of the Aztecs, based on these assumptions, is shown in table 3.7.

Cannibalism was not a significant adjunct to the Aztec diet, on the basis of Dornstreich and Morren's (1974) criterion of 5–10 percent of dietary protein, except for total body consumption in Tenochtitlan, which was not the usual case. If Harner and Harris were right that cannibalism was a Darwinian adaptation to a lack of animal protein, then it was a singularly unsuccessful adaptation, especially since it was available only to the nobility, a minority (25 percent) of the population. Because Harris (1979: 337) disparaged the potential contribution of wild game to the nutrition of *all* the inhabitants of the Basin of Mexico, let us calculate the contribution of cannibalism if human meat were divided among all the inhabitants rather than only the nobility (table 3.8). Table 3.8 demonstrates graphically how illogical it is to regard cannibalism as an adaptive source of protein. At the maximum, even Tenochtitlan nobles consuming the whole bodies of sacrifices would receive only about one kilogram of protein per year (2.85 g per day), equivalent to about 4 (2-lb) fishes or 3 (4-lb) ducks a year, or about half an ounce of duck or a third of an ounce of fish a day. At the lower limit for Tenochtitlan (total population sharing only extremities), cannibalism would provide the equivalent of a quarter of a duck or half a fish per person per year, on the order of four hundredths of an ounce of fish or duck per day.

Although Harris (1979: 337–339) derided the significance of wild animal food in the diet of the Aztecs, his own calculations showed that the contribution of wild game and insects would have greatly exceeded that of cannibalism. Using Gibson (1964: 340, 343), Harris (1979: 339) calculated that at the time of the Conquest three million fishes and one million ducks per year were caught in the lakes of the Basin of Mexico, giving every Aztec (notice that he includes *all* the population and not just the nobility) 0.6 g protein from wild game, 0.12 g from fish, and 1.0 g from duck, which with 1.0 g from insects yield a total of 2.7 g protein per day. This is ten times as much as the average resident of Tenochtitlán would have derived daily from cannibalism (0.25 g) if only extremities were

Table 3.7. Protein Contribution of Cannibalism to Diet of Aztec Nobles

| LOCATION | PROTEIN NEED[a] | AVAILABLE PROTEIN[b] | | NEED SATISFIED | |
	kg/yr	Whole body	Extremities[c]	Whole body	Extremities
Tenochtitlan	1.2×10^6	7.8×10^4	2.72×10^4	6.5	2.3
Central Mexico	9.7×10^7	1.3×10^6	4.5×10^5	1.3	0.46

Source: Ortiz de Montellano 1978. Copyright 1978 by the AAAS.
[a]Assuming 75,000 consumers in Tenochtitlan and 6.26 million in Central Mexico.
[b]Assuming 15,000 annual sacrifices in Tenochtitlan and 250,000 in Central Mexico.
[c]Assumed to be 35% of body weight.

Table 3.8. *Annual Protein Contribution of Cannibalism to Aztec Nutrition*

LOCALITY	AVAILABLE PROTEIN kg	NEED SATISFIED %		PROTEIN/PERSON kg		DAILY PROTEIN/PERSON g	
		Nobles	All	Nobles	All	Nobles	All
Whole body							
Tenochtitlan	7.8×10^4	6.5	1.6	1.04	0.26	2.85	0.71
Central Mexico	1.3×10^6	1.3	0.33	0.21	0.05	0.58	0.14
Extremities only							
Tenochtitlan	2.72×10^4	2.3	0.58	0.36	0.09	0.99	0.25
Central Mexico	4.5×10^5	0.46	0.12	0.07	0.02	0.20	0.05

consumed, and fifty-four times as much as the average inhabitant of Central Mexico would have derived (0.05 g) from this "adaptive" practice.

Ortiz de Montellano (1978) also showed that the timing of sacrifices and cannibalism was inconsistent with the Harris-Harner thesis. Sacrifices were not performed during periods of the agricultural cycle when scarcity of food might be expected. Rather, sacrifices and cannibalism were at their maximum during harvest. This negates Harner's hypothesis and tends to support the traditional explanation of Aztec sacrifice and cannibalism as ritual events giving thanks to the gods for fertility and good crops.

= *Agricultural Systems*

A large population was maintained before the Conquest by a very productive agricultural system based on multicropping, chinampas, irrigation, terracing, and the cultivation of a group of superior plants. This achievement is similar to that of China, where a fourth of the world's population is fed on a relatively small cultivated area without the use of chemicals or machines. Anderson (1988: ix) attributed the Chinese capability to an agricultural system that is not only productive but sustainable, based on highly productive crop varieties, recycling of nutrients, efficient use of water resources, and skilled intensive labor. Up to now, "traditional peasant" systems have usually been considered relatively unproductive, with a very large labor input per unit of crop output. As Truman (n.d.) stated:

> What agro-ecologists and biologists are discovering around the world is that these "tradition-bound" systems are not only very productive but are often built upon biologically sophisticated principles that have only recently been addressed by modern day researchers.

One of these "biologically sophisticated principles" is multicropping, or intercropping, in which several different species are planted simultaneously, rather than limiting farming to one exclusive crop. This practice has the effect of maximizing production and diminishing the risk of complete crop failure due to poor weather or pest damage (Cox 1978). The principal food triad of Mexico, since plants were first domesticated, has been corn, beans, and squash, often planted together. Experiments have shown that such planting significantly increases the maize yield. Beans are legumes that can fix nitrogen biologically and make it available in the soil as a fertilizer. Squash helps to control weeds by creating a dense shade that allows very little light to penetrate, and chemicals washed from its leaves enter the soil to act as herbicides (Truman n.d.). Intercropping of beans and corn also reduces the damage caused by the corn-ear worm and increases the crop yield by fifty percent (Downs 1984). Intercropping has been recommended as a way to decrease the need for pesticides and fertilizer in modern agriculture (Horwith 1985). Bye (1981) described intercropping in which amaranth is planted simultaneously with corn, several inches above the level of the corn seed. The amaranth grows faster than corn, so its seedlings are available for harvesting as edible greens before they interfere with the corn crop. About 100 g of edible seedlings can be obtained per week for several weeks from a plot of 1–4 m². The seedlings form an important part of the diet:

> Being subsistence agriculturalists, the Tarahumara depend on an annual diet cycle based upon maize, bean, cucurbit and chile which are consumed from fresh plants in August through October and from stored, dried forms in October through May. Often times the stored cultivated food supplies are limited from April through July. During the latter period, the diet is augmented by hunted and gathered resources such as fish, wild greens, roots, bulbs, and "hearts" of maguey (*Agave* ssp.). It is during this period that quelites [edible greens] from the cultivated fields dominate the diet. May–June period also marks the end of the dry season and the beginning of the rainy period and the start of the annual growing season.

> The weeds can also provide food after the initial growing period. July and August may be frequented by severe hail storms which destroy the young maize plants. Also, animal pests such as crows and insects can destroy portions of the maize crop at different stages. The tender apices of the older weed plants as well as the late emerging seedlings can be collected and consumed. The quelites represent a living emerging food reserve (Bye 1981: 114)

The quality of amaranth greens is also significant. Bye pointed out that the Tarahumara diet is liable to be deficient in calcium, vitamin A, thiamine, riboflavin, and vitamin C, several of which are compensable by a daily supplement of 100 g of amaranth to meet U.S. RDA standards. Multiple cropping has also been shown to be a very effective and productive way to manage forests in order to supplement staple crop farming with a minimum of labor input (Alcorn 1984).

Chinampa cultivation was another sophisticated agricultural technique. Chinampa planting surfaces are created by lifting masses of soil and aquatic vegetation and consolidating them high enough above the water of a shallow lake so that seeds can germinate properly. The result, typically, is a long ridged field, two to four meters wide and twenty to forty meters long, surrounded by ditches navigable by canoe. The fertility of a chinampa can be continuously renewed with mud sediments and aquatic vegetation scooped from the lake surrounding the field, supplemented by household refuse. There is some indication that, in late Aztec times, large quantities of bat dung and human excrement also were used (Parsons et al. 1983: 51). Chinampas were not affected by droughts and produced up to seven crops a year, starting with seedlings including two corn crops (Coe 1964). Annual corn yields can be conservatively estimated at 3,000 kg/ha and, as described above, some 12,500 ha were in cultivation by 1519.

Most chinampas were built after the great famine of 1450. Their yields are usually quoted in terms of maize because most of the data has been obtained for this crop, but chinampas were probably interplanted with beans and squash. Recent evidence (Chen 1987; Gómez-Pompa 1987) showed that the Classical Maya sustained population densities of 116–154/km^2 in a tropical forest ecosystem, compared to a density of 10/km^2 today in rural Maya areas. Such large populations could not have been maintained by swidden agriculture, which ordinarily supports a population density of only 25/km^2. Archeological evidence for the existence of Mayan chinampa fields suggests that they made possible the high population density during the Classic Period. An important point to be made is that, through chinampa farming and forest management, the Maya not only sustained a large population but also preserved the tropical forest (Chen 1987; Gomez-Pompa 1987). By contrast, modern agricultural techniques are destroying a significant percentage of the world's tropical forest each year, without supporting a population comparable to that of the Maya.

A third agricultural technique is *metepantli*, a type of terracing that predates the arrival of the Europeans in the New World and is today still a viable means of farming seasonally dry, sloping land (Patrick 1977: 1). Characteristic metepantli features are a retainer wall, held securely in place by the root systems of *maguey* (*Agave sp.*) plants located on its top, and an upslope inclined planting surface held in place by the retainer wall. We shall see that maguey plants served a number of purposes, including the provision of food and drink. The metepantli fields in Patrick's study had a concave surface whose lowest point was usually located within the curve, instead of along the terrace's downslope edge. Such contouring diminishes rainfall runoff and thus helps to prevent erosion. Its role is even more important in the transition period of rainfall, when newly planted shoots of maize, and later beans, are fragile and require a constant and adequate level of moisture. During this phase, from March to mid-May, the ability of the concave metepantli to trap and retain whatever rain falls, and to conserve moisture, is crucial to good yields (Patrick 1977: 119).

The current average yield of dry corn in metepantli fields is 4,018 kg/ha (Patrick 1977: 200) and reflects the use of manure fertilizer. Before the Conquest, these fields were probably allowed to lie fallow in alternate years, which would halve the land-use factor. Southern Tlaxcala, the region studied by Patrick, is today the most densely populated rural area in Mexico beyond the Federal District and the rural fringes of Guadalajara and Monterrey, with a density of over one hundred persons per square kilometer in 1970 (Patrick 1977: 169). Seventy-one percent of the population lives in an area where metepantli agriculture is practiced. This nonirrigated land still produces an exportable surplus. As Patrick (1977: 210) concluded:

> the figures suggested here, if applied to other communities on the slope of La Malinche, lend support to the conviction that metepantli terraced lands in the Southern Tlaxcala region, in conjunction obviously with socio-technical variables of crop production, can provide caloric sustenance to its rather densely settled occupants at a level far above that of subsistence. Surplus maize yields are available annually to practitioners of metepantli agriculture, which when sold provide sources of income for the purchase of other life necessities.

The Aztec Diet:
Food Sources
and Their
Nutritional Value

= Cultivated Food Plants

The demographic success of the Aztecs was due not only to the efficient agricultural techniques discussed in chapter 3 but also to the sowing of very productive food plants adapted to the arid climate. New World food plants were inherently superior to staples of the Old World; in fact, their introduction into the Old World fueled a population explosion (McNeill 1976: 231). Corn will grow in areas too dry for rice and too wet for wheat, and its average yield per unit of land area is roughly double that of wheat. Few other plants produce as much carbohydrate and sugar in as short a growing season as does corn (Crosby 1972: 171). Hernández (1959: vol. 1, 288) sang its praises in the sixteenth century:

> I praise it greatly, and I don't understand how the Spanish . . . have not adopted it for their use nor taken it to their lands and cultivated this species of grain. . . . It is easily cultivated and produces abundantly and most assuredly in almost any soil and which is little subject to the harm of drought and other rigors of the sky and earth. Through it they might perhaps free themselves from hunger and of the numberless evils that derive from it.

Dubos (1965: 284) attributed the extremely fast population growth in China, from under 64 million in 1578 to 108 million by 1661, to the introduction of easily grown and very high yielding New World plants: corn (1550), sweet potato (1580), and peanuts somewhat later. In Spain and Italy also, where cultivation of maize became widespread, populations soared—nearly doubling in Spain in the eighteenth century (Farb and Armelagos 1980).

New World food plants not only increased the quantitative yield of calories per hectare in the Old World but also improved the quality of the diet.

> Chili peppers and tomatoes, for instance, supplied a rich vi-
> tamin source whose importance in the diets of Mediterranean
> and Indian populations in modern times is very great in-
> deed. . . . As these foods entered into widespread use among
> rich and poor alike, one can be sure that a more adequate diet
> became available to the Indian and Mediterranean peoples,
> and health levels presumably reflected this fact. (McNeill 1976:
> 231)

The high productivity of corn (and a second staple, amaranth) is due to inherent biological characteristics not special agricultural techniques. These plants are intrinsically superior in thermodynamic efficiency, defined as the ratio of food energy output to solar energy input. The underlying biochemical process is the photosynthetic conversion of carbon dioxide and water into glucose and oxygen, energized by the absorption of photons of light.

As the second law of thermodynamics states, no conversion of energy from one form to another is 100 percent efficient; solar energy cannot be converted into food energy without some waste. The maximum theoretical efficiency of photosynthesis is 12.3 percent,[1] that is, production by a plant of 12.3 kcal of food energy for every 100 kcal of solar energy it receives. (1 Cal, the unit used for food energy, equals 1,000 cal, or 1 kcal, where 1 cal is the unit used for physical energy.) This efficiency is based on photosynthesis in aquatic plants (for example, unicellular algae) and the assumption that no other wasteful processes occur. The maximum theoretical efficiency is lower for land plants because their leaves overlap, creating shade, and also reflect light away, thus preventing absorption of all the incident radiation (the combined action is called the canopy effect). Another factor is plant respiration. Plant cells use energy when they are not photosynthesizing, for example, at night, or in plant tissues that are shaded or not green. These factors place a 6.6 percent limit on the maximum theoretical efficiency for the conversion of solar radiation into food energy by land plants.[2] This translates into the production of the equivalent of 72 g of plant mass/m² per day, or 263 metric tons/ha per year in an area with abundant sunlight.[3] These are the maximum *theoretical* yields determined by the physics and chemistry of the process and the second law of thermodynamics. Actual maximum yields are only about half the theoretical limit.

A further complication is that plants carry out photosynthesis by two different mechanisms, called the C_3 and C_4 pathways. Species that use the C_4 pathway are inherently more efficient than those using the C_3 pathway (Zelitch 1979).[4] Thus, under identical conditions of sunlight, water, and soil, C_4 plants produce much more than C_3 plants. Two of the staple Aztec foods (corn and amaranth) belong to the more efficient C_4 group, while wheat, rice, and soya, for example, are in the C_3 category. Zelitch (1979) stated that average annual yields per acre in 1976 were 4,872 lb of corn and only 1,800 lb of wheat. We can now see the reason for Hernández's enthusiasm and the effect corn had when introduced into the Old World. Superior productivity is also a factor in the usefulness of tecuitlatl (*Spirulina geitlerii*), discussed below.

Of particular relevance to a consideration of Aztec foods is the nutritional value of the staple diet of Mesoamerica—corn, beans, and squash, supplemented by chilies and tomatoes. This triad has been the basis of the Aztec diet since antiquity, and the addition of chili and tomato as condiments covered most culinary situations. The importance of maize to the Aztecs is reflected by the many names they gave it; for example, each growth stage had a specific name. Ethnobotanists have shown that plants of great cultural significance tend to be overdifferentiated (Berlin, Breedlove, and Raven 1966). Maize was referred to as *tonacatl*, "our flesh," and women treated it reverently before cooking it (Sahagún 1950–1969: books 4–5, 184). The basic group was also supplemented by an extraordinary variety of other foods, to be discussed later.

Araya, Flores, and Arroyave (1981) postulated that adequate nutrition can be achieved by a diet deriving 30 percent of its protein from beans and 70 percent from corn, a proportion close to Sanders's assumption that 80 percent of the calories in the Aztec diet came from corn. Table 4.1 shows the nutritional values of such a diet with the addition of squash, chilies, and tomatoes, which were also usu-

Table 4.1. Nutritive Values of Model Aztec Diets

	CAL	PROTEIN	CA	P	VIT. A	THIAMINE	RIBOFLAVIN	NIACIN	VIT. C	FE
Diet 1										
Corn (500 g)	1,790	42	55	605	0.75	2.4	0.5	9.5	—	7.5
Beans (100 g)	343	22.7	134	415	0.008	0.47	0.15	2.09	1	7.1
Squash (50 g)	10	0.6	14	15	0.06	0.2	0.05	0.5	11	0.2
Chili (20 g)	23	0.93	7.3	19.5	1.62	0.06	0.09	1.1	93	0.3
Tomato (50 g)	11	0.6	7.0	14	0.13	0.03	0.02	0.4	12	0.3
Total	2,177	66.83	217	1,069	2.57	3.16	0.81	13.6	117	15.4
Diet 2										
Corn (300 g)	1,074	25.2	33	363	0.45	1.44	0.3	5.7	—	4.5
Amaranth (200 g)	782	30.6	980	910	—	0.28	0.64	2.0	9	34.8
Beans (100 g)	343	22.7	134	415	0.008	0.47	0.15	2.09	1	7.1
Squash (50 g)	10	0.6	14	15	0.06	0.2	0.05	0.5	11	0.2
Chili (20 g)	23	0.93	7.3	19.5	1.62	0.06	0.09	1.1	93	0.3
Tomato (50 g)	11	0.6	7.0	14	0.13	0.03	0.02	0.4	12	0.3
Total	2,243	80.63	1,175	1,737	2.27	2.48	1.25	11.79	126	46.4
FAO-WHO	2,200	45	800	800	1.0	1.2	1.8	20	45	

Note: Units are g for protein and International Units (IU) for vitamin A. All others are mg.

ally eaten. Almost every source refers to the universal use of chili: "They have one—like pepper—as a condiment which they call chili, and they never eat anything without it" (Anonymous Conqueror 1972: 37). Amaranth was used as an alternate to corn and grown in large amounts, as demonstrated by the fact that tribute to Tenochtitlan included an amount of amaranth equal to half the amount of corn. Table 4.1 also shows the nutritional values of a second model diet in which 200 g of amaranth replaces 200 g of corn.

Diet 1 is low in calcium, riboflavin, and niacin, but the way in which corn was prepared helped to compensate for the deficiency. Traditionally, dried corn was heated almost to the boiling point in a 5-percent solution of lime (CaO) in water for 30–50 minutes. The mixture was cooled and the corn washed thoroughly and drained. The corn was then ground into a dough called *masa* and cooked on a griddle to make tortillas. Alkaline cooking enhanced the protein quality of the corn by beneficially altering the relative amounts of different essential amino acids and by making both niacin and its precursor, tryptophan, more available to absorption by the body (Katz, Hediger, and Valleroy 1974). Lime soaking also increases the calcium content of corn from 11 to 158 mg/100 g. Katz, Hediger, and Valleroy pointed out that populations whose diets are composed primarily of maize would be expected to develop pellagra as a consequence of niacin deficiency, unless other dietary constituents contribute niacin, its precursor, or both. In other corn-consuming areas, such as South Africa, India, and the southern United States earlier in this century, where corn is not a native dietary item, pellagra is a major disease. In Mesoamerica, pellagra is and was rare, due partly to alkaline processing. Also, the deficiency of both niacin and lysine in corn is ameliorated by pairing corn with either beans or amaranth, both high in lysine (see appendix A).

Another form of preparation that enhances corn's nutritional value is *pozol,* a maize dough fermented by a mixture of yeasts and other microorganisms, which, after dilution in water, can be drunk raw. Cravioto et al. (1955) show that pozol has more protein, niacin, riboflavin, lysine, tryptophan, and other nutrients than ordinary maize, and its protein is also of higher quality.

═ *Tecuitlatl (Spirulina geitlerii)*

One of the unusual Aztec foods encountered by the Spanish was a green substance called tecuitlatl, collected from salty Lake

Figure 4.1. Gathering the algae *tecuitlatl (Spirulina geitlerii)* from the surface of Lake Texcoco in the sixteenth century. (Sahagún 1906. *Florentine Codex*: book 11)

Texcoco (fig. 4.1). It was sold in the market and eaten with maize or a sauce made of chili peppers and tomatoes. Tecuitlatl was briefly described by Díaz del Castillo (1968: vol. 1, 279), one of the first Spaniards to enter Tenochtitlan: "then there were fishmongers and others who sold little loaves which they make out of a sort of slime which they gather from the great lake, which they thicken and they make loaves of it which taste like cheese." Sahagún (1956: vol. 3, 263; 1950–1969: book 11, 65) also mentioned that it was gathered in the lake and that it was edible. Much more extensive descriptions were given by two other sixteenth century chroniclers. A Franciscan friar, Toribio de Benavente (called Motolinía, "the poor one"), spent many years among the Aztecs and, about 1541, wrote:

> There breeds upon the water of the Lake of Mexico a sort of very fine slime, and at a certain time of the year when it is the thickest, the Indians gather them with very fine nets until their *acales* or boats are full. On shore they make a very smooth plot two or three brazas [3.4–5.1 m] long and a little less wide on the earth or on very fine sand. They throw it [the tecuitlatl] down to dry until it makes a loaf two *dedos* [3.6 cm] thick. A few days later, it dries to the thickness of a used ducat. The Indians cut this loaf into wide bricks and eat a lot of it and think it good. This merchandise is carried by all the merchants of the land as cheese is among us. Those of us, who share the tastes of the Indians, find it very tasty. It has a salty flavor. I think particularly that this substance is the bait, which brings great multitudes of birds to the Mexican lagoon. There are so many, that in many parts, it looked like a solid lake made up of birds. This happens in winter and the Indians harvest many of these birds. (Motolinía 1971)

Hernández (1959: vol. 2, 408–409) classified tecuitlatl as a mineral and was less enthusiastic about the taste:

> Tecuitlatl, which resembles slime, sprouts in certain places in the Mexican basin and rises to the surface from which it is swept with nets or is removed with shovels. Once extracted and slightly sun dried, the Indians shape it into small loaves. It is again placed on fresh leaves until it is completely dry and then it keeps like cheese for only one year. It is eaten as necessary with roasted corn or with the common tortillas of the Indians. Each source spring of this slime is owned privately and sometimes yields a profit of a thousand gold ducats yearly. It tastes like cheese (and so it is called by the Spaniards), but it is less pleasant with a certain muddy smell. When it is fresh, it is blue or green, when it is old it is the color of slime, a blackish green. It is edible in only small amounts and in place of salt or as a flavor

for corn. The tortillas that are made from it are a poor and rustic food, which is proven by the fact that the Spaniards, who never turn down anything that pleases the palate, particularly here, have never eaten them.

Motolinía, who spent much more time with the Indians and was personally acquainted with their customs, is a more trustworthy reporter than Hernández, who was repelled by many Aztec food habits.

Reports of the use of tecuitlatl decreased with time as the lakes were drained and lost their importance, and the identity of the substance was lost. The breakthrough in identification came not from Mexico but from a Belgian scientist, Jean Leonard, who found that blue cakes called *dihe,* eaten by people along the shores of Lake Chad in Central West Africa, were composed of the algae *Spirulina geitlerii* (Furst 1978). The algae are gathered from the lake surface, dried into loaves by the sun, and eaten with a sauce made of tomatoes and chili peppers (note the resemblance to Motolinía's description). Apparently, all these ingredients were introduced to Chad after the conquest of Mexico.

Spirulina turns out to be a remarkable food. It is 70 percent protein with an essential amino acid assortment very similar to that of egg, considered the ideal combination. It is also rich in vitamins and minerals, especially phosphorous, thiamine, riboflavin, and niacin (see appendix A), which makes it an excellent supplement to corn. In contrast to other algae such as *chlorella,* which has also been proposed as animal feed but is difficult to digest because of celluloselike cell walls, *Spirulina* is a prokaryote. Its cell wall is mainly easily digested noncellulosic material. It is also relatively low in nucleic acid, which if consumed in excess can cause gout (Ciferri 1981). The virtues of *Spirulina* were verified in long-term feeding trials on humans in France and Mexico that included up to 100 g per day in a diet without ill effects (Durand-Chastel: 1980).

The testimony of Motolinía and Hernández about the abundance of *Spirulina* is credible since, as indicated above, the potential yield of algae is twice that of any land plant. Studies show that *Spirulina,* under optimal conditions, can yield up to 20 g/m^2 dry weight per day. Table 4.2 compares the yields of some traditional crops to that of *Spirulina.*

We can calculate how much *Spirulina* the Aztecs needed for food and how much lake surface would have been required to

Table 4.2. *Comparative Yields of Traditional Crops and* Spirulina

CROP	YIELD, TON/HA YR	
	Dry weight	*Crude protein*
Wheat	4	0.5
Maize	7	1
Soya	6	2.4
Spirulina	50	35

Source: Ciferri 1981. This appeared in *New Scientist* magazine, London, the weekly review of science and technology.

grow it. Assumed populations of 300,000 for Tenochtitlán and 1,200,000 for the Basin of Mexico would have needed 0.25 percent and 0.95 percent, respectively, of the surface area of Lake Texcoco to provide them with their complete protein requirements for a year.[5] Motolinía's comment about the number of birds attracted to feed on the algae is evidence of the abundance of *Spirulina* being produced on the lake. Also the quantities sold and consumed must have been fairly large to bring a marketer profits of up to 1,000 gold ducats a year.

Another advantage of *Spirulina* is that it grows in salty lakes and so does not compete with or displace other potential food sources. In effect, it converts a liability into a resource. It is droughtproof, because it does not depend on rainfall, and stable for up to one year after drying, so it can be prepared when abundant and stored for use in times of scarcity. Santley and Rose (1979) were so impressed with *Spirulina's* dietary potential that they credit it with the high population growth rate in the late Aztec period. *Spirulina* is currently being reinvestigated as a possible solution to malnutrition in the Third World, particularly in areas subject to severe drought (Durand-Chastel 1980).

▭ Desert Plants

Particularly useful as food resources, given the semidesert climate of the Basin of Mexico, are plants adapted to low rainfall.

They are not only productive in normal years but essential in times of drought. Three of the most important are amaranth (*huauhtli Amaranthus sp.*), mesquite (*mizquitl, Prosopis sp.*), and maguey (*metl, Agave* sp.). As mentioned above, large amounts of amaranth were grown and delivered annually to Tenochtitlan. Amaranth grows very rapidly, particularly under conditions of high temperature and dry soil, and, because it belongs to the C_4 group of plants, is extraordinarily productive. It can flourish in environments ranging from the true tropics to deserts and from sea level to an altitude of 3,000 meters (National Research Council 1984: 14). The native habitats of the grain amaranth (*A. Cruentus, A. hypochondriacus*) in North America are the deserts of Arizona and the northwest and high central plains of Mexico. These species are useful as both grain and leafy vegetable. Some of their growing characteristics were described in a U.S. Department of Agriculture bulletin:

> Among the vegetables of the tropics, few are as easy to grow as the amaranths. Starting from tiny seed, these species can produce delectable spinach-like greens in 5 weeks or less [We have seen how this property was used in intercropping with corn.], can continue to produce a crop of edible leaves weekly for up to 6 months, and will then produce a crop of seeds to guarantee their survival. In favorable locations, they can reseed themselves automatically and thus continue to produce a useful crop almost without attention. . . . In the tropics, amaranths can produce year-round. For little effort, they afford a nutritious dish with abundant provitamin A, a vitamin particularly necessary in the tropics for eye health. Amaranths also produce protein efficiently. (Martin and Telek: 1979)

The grain contains 16–18 percent protein, compared with 14 percent or less for wheat and other grains, and its protein has an excellent balance of amino acids (see appendix A). It is unusually rich in lysine, the amino acid in which other grains, such as corn, are usually deficient. The protein quality score of tested amaranths ranges from 75 to 87, higher than cow's milk (Hindley 1979). Its high lysine content means that it can substitute for beans as a compensatory companion of corn. Amaranth leaves are also high in protein (6.2 percent) and contain a large amount of vitamin A (see appendix A).

In spite of amaranth's great nutritive potential, its cultivation declined precipitously after the Conquest because of its role in Aztec religion and ritual. Amaranth dough was used to make images of the gods, eaten in certain festivals as a form of communion (Sahagún 1950–1969: book 2, 112–116; Durán 1971: 85). This practice was so common that amaranth dough was referred to as "bones of God" (Durán 1971: 245). Amaranth-dough images of mountains, representing tlaloque, were used in curing rituals during the month of Tepeilhuitl (Sahagún 1950–1969: book 2, 256; Durán 1971: 256). Amaranth and corn were often identified as the only food that could be eaten during certain ritual periods. The ritual use of amaranth dough extended to domestic ceremonies. When babies were ceremonially bathed and named, four days after birth, they were also given small replicas of their future tools made from amaranth dough. Boys were given a shield, bow, and arrows, and girls were given domestic implements (Sahagún 1950–1969: book 6, 201). The Spanish conquerors, determined to convert the Indians to Christianity and to stamp out the old "heretic" religion, with which the use of amaranth was inextricably intertwined, banned its cultivation.

Mesquite, another nutritive desert-adapted plant eaten by the Aztecs, featured in the Aztec mythology of the evolution of food during repeated creations and destructions of the world (see chap. 2). In general, the particular food eaten by people created in each era slowly evolved to culminate in "the perfect food," corn. Moreno de los Arcos (1967) tried to reconcile the various versions of this myth. His reconstruction has the following sequence: Era I—acorns, Era II—mesquite, Era III—*cincocopi* ("almost corn"), Era IV—*acicintli* (possibly "water corn"), and Era V—corn. This is a case of myth coinciding with archaeological fact. We know that Central Mexico was peopled by recurring southward migrations of hunting-gathering groups, which settled down and assimilated the pre-existing cultural heritage. Mesquite has been a prime resource for hunter-gatherers in the region since antiquity. Analyses of coprolites (Callen 1973) from the Ocampo caves in Tamaulipas and the Tehuacan caves in Puebla show that *Prosopis* (mesquite) was an important part of the diet throughout a span of several thousand years, second only to *Agave* (maguey), to be discussed next. This dependence continues today. Mesquite pods and seeds, together with species of *Agave* and *Amaranthus*, are among the most important food staples of the Seri, hunting and gathering Indians of Sonora, Mexico (Felger and Moser 1976).

Felker and Bandurski (1979) profiled the "ideal" characteristics of a food tree: 1) growth conditions that cause minimal soil and nutrient loss, 2) little or no need for irrigation, 3) high yield, 4) ability to fix nitrogen, and 5) production of large amounts of high quality protein. Mesquite fits this profile extremely well: 1) It is a perennial tree that requires no tilling and thus preserves the ground cover and minimizes soil loss. 2) It is extremely drought resistant since its root system commonly extends 20 m (so much as 80 m, if needed) down to the water table. 3) Yields are exceptional, even under very dry conditions. The mesquite species *P. tamarugo* yields 10,000 kg/ha of pods in an area where rain falls once every several years. By comparison, millet and peanut yield about 1,000 kg/ha in areas with 600 mm of annual rainfall (Felker and Bandurski 1979). 4) Mesquite is a legume and therefore capable of fixing nitrogen from the atmosphere. 5) Mesquite pods and enclosed seeds are both nutritious foods. Whole dried pods contain 15.8 g protein, 3.0 g fat, and 50.8 g carbohydrate per 100 g, a combined energy value of 294 cal (Gupta, Gandhi, and Tandon 1974). The seeds are even richer in protein, up to 69 percent for the seed embryo and 40 percent for the whole seed, and the quality of the protein is also high (see appendix A). The absence of protein-calorie malnutrition in several villages in India, a malnutrition expected because of semidesert location and poor socioeconomic status, is attributed to the consumption of mesquite. Unripe pods were used to make curry and pickles, and mesquite was considered particularly beneficial for young children

who, while playing about in the fields, pick up ripe pods and eat the pulp and outer covering with relish, discarding the seeds. This soft portion is a fairly rich food, providing 10.8 g/100 g protein and 252 kcal/100 g. (Gupta, Gandhi, and Tandon 1974)

Metl (maguey in Spanish, *Agave* sp.) is a third desert plant consumed extensively in Mexico. Quids, the chewed fibrous remains of roasted maguey leaves, were found in the Tehuacan caves. A common cooking method involved a type of barbecue consisting of a pit dug in the earth, lined with heated rocks, and then filled with the maguey leaves and covered (Coe 1985). Maguey, which often represented the most abundant food residue in the Puebla and Tehuacan caves (Callen 1973), has served many purposes besides food from antiquity to the present, for example:

As a whole it can be used as a fuel or to fence fields. Its shoots can be used as wood and its leaves as roofing materials, as plates or platters, to make paper, to make cord with which they make shoes, cloth and all kinds of clothes, which among us are made of linen, or cotton or similar fabrics. . . . From the sap, which flows into the central cavity when the shoots or more tender leaves are cut with a flint, they make wines, honey, vinegar and sugar. Sometimes a single plant will produce fifty amphoras [1,200 liters] of this juice. . . . From the root, they also make very strong ropes which are useful for many things. The thicker parts of the leaves as well as the trunk, cooked underground (a cooking method which the Chichimec Indians call *barbacóa* [barbecue today]), are good to eat and taste like cidra [a citruslike fruit] with sugar. . . . The plant reproduces by shoots, which grow around the mother plant in any soil but primarily in a fertile, cold one. This plant, by itself, could easily furnish all that is needed for a simple, frugal life since it is not harmed by storms, the rigors of the weather, nor does it wither in drought. There is nothing which gives a higher return. (Hernández 1959: vol. 1, 348–349)

The plant's versatility continues. Patrick (1977: 94) listed two pages of uses of the maguey, which have changed very little over the centuries. In Patrick's words:

Maguey plants grow best on arid lands at high altitudes. . . . The magueys are reproduced from shoots that germinate at the bottom of the mature plants. . . . It takes from eight to twelve or more years before the plants can be exploited for pulque [a fermented alcoholic drink], those growing in the higher altitudes taking longer but yielding finer juices. . . . If the plants are not cut into on time, a tall stalk grows from the center, bearing quiotes or yellow flowers. They are good to eat when fried, but the plant is spoiled for pulque. . . . A maguey yields a few quarts of juice a day for some six months. After that it is finished for pulque, although it is still useful. These plants are managed so that each year new shoots are planted to compensate for those that are consumed for food and drink. (Patrick 1977: 112)

Fermented "honey water," called *octli* by the Aztecs and *pulque* in Spanish, was the principal alcoholic drink. As discussed in chapter 2, pulque was deified as the god Ome Tochtli (2-Rabbit). The maguey plant itself (and fertility) was identified with the goddess Mayahuel, and other gods, collectively called the *tzentzon totochtin* ("400 rabbits"), represented some of the behavioral quirks of people who were drunk. They were also often patron gods of villages, for example, Tepoztlan whose patron god was Tepoztecatl, one of the "400." The fact that pulque gods were fertility-agricultural deities and members of the Tlaloc complex indicates the importance and usefulness of the maguey.

Pulque was also a food resource. Fermentation adds protein, so that a liter of pulque has 204 cal, 4.4 g of protein, 3.9 mg of niacin, and a whopping 62 mg of vitamin C. Although Aztec culture had many prohibitions against getting drunk, pulque was considered healthful in moderation. Motolinía (1971: 363), after fulminating against the evils of drunkenness, stated that "in reality, drunk in moderation it is healthy and strengthening. All medicines which they [the Indians] administer orally are given to the patients with this wine in their cup; there they place whatever is restorative for each illness." In desert areas, where potable water is scarce, pulque becomes a crucial source of liquid to the inhabitants, for example, the Otomi Indians.

The Otomi Indians in Mexico are a classic example of how well a population can survive on unusual foods in an inhospitable environment (Anderson et al. 1946). According to the investigators who studied them, the Otomi lived in the poorest region of Mexico, the Valley of the Mezquital. It is about 2,000 m high, exceedingly arid, and essentially unsuitable for agriculture. Average annual rainfall is 440 mm, and only on an average of fifty-two days does the rainfall exceed 0.1 mm. During the study, water was so scarce that the Otomi seldom washed either their clothing or their bodies. Sanitary facilities were practically nonexistent, and medical care was meager. Each village had one or two deep wells, but the water was unsafe to drink. Pulque was drunk in large quantities, about one to two liters daily per person, and often replaced water entirely. Since the Otomi ate very few of the foods, such as meat, dairy products, fruits, and vegetables, commonly considered essential for good nutrition, they were expected to suffer from general malnutrition and vitamin and mineral deficiencies. Their usual diets were analyzed, and they were given physical examinations, particularly for symptoms of deficiency diseases. Blood samples were analyzed for protein and vitamin content. To the surprise of the investigators, the diets were found only moderately below U.S. RDA standards for calories and protein at the time [which we now know were too high], but intakes of vitamin A, thiamine, ascorbic acid, calcium, and phosphorus were satisfactory.[6] Table 4.3 shows the analysis of the Otomi diet.

Anderson et al. (1946) described in detail the diet that produced such surprising results in a difficult ecological situation:

Table 4.3. Nutritive Values of Otomi Diet

	CALORIES	PROTEIN	CA	FE	VIT. A	THIAMINE	RIBOFLAVIN	NIACIN	VIT. C
RDA	2,430	64	960	11.4	4331	1.19	1.66	11.9	68
Otomi diet	1,706	51	820	23.5	5498	1.61	0.69	9.4	96
% RDA	70	80	85	206	127	135	42	79	141

Source: Anderson et al. 1946. Reprinted by permission of the *American Journal of Public Health*.
Note: Units are g for protein, International Units (IU) for vitamin A. All others are mg.

As in other parts of Mexico, the basic foods here are corn (almost invariably eaten as tortillas), dried beans, and chili peppers. Supplementary foods are chiefly those readily available locally, the most important of which is the fermented maguey (agave) juice, pulque. Meat is eaten in very small quantities and is usually from sheep or goats, since these are the animals which thrive best in this arid region. Often the blood alone is eaten and the meat is sold to buy cheaper foods. The same is usually true of the small amounts of milk, eggs, and poultry produced. Rabbits are practically the only wild game eaten. Beans are relatively expensive and are used in lesser quantities than in many other parts of Mexico. Chili consumption is large. Small amounts of onion, garlic, "tomate" (a plant of the ground cherry family), and less commonly, tomatoes (jitomates) are used for flavoring purposes. . . .

Almost every conceivable edible plant, including many of the cacti, are used as foods. Many grow without cultivation, during the rainy season, and by most people would be considered as weeds. A variety of worms and insects are also eaten with relish. By these means a diet of considerable variety is attained. Some of the "weeds" are surprisingly nutritious, [see appendix A] e.g., the prolific malva [mallow, *Malva parviflora*] is an exceptionally good source of vitamins A and C and calcium and iron. Others of some importance in the diet include hediondilla [lamb's quarters, *Chenopodium sp.*]; tunas and nopales (prickly pear cactus fruit and leaves respectively) [*Opuntia sp.*]; flowers of maguey (century plant), garambullo (a myrtilocactus) and yuca [yucca, *Yucca sp.*]; verdolagas (purslane [*Portulaca sp.*]); quelites (pigweed [*Amaranthus sp.*]); xocollol (wood or sheep sorrell [*Rumex sp.*]); flor y hojas de nabo (flowers and leaves of a variety of wild mustard); lengua de vaca (a variety of dock [*Rumex sp.*]); and endive (sow thistle [*Sonchus oleraceus*]). Fruits, except for those of several of the cacti are almost never eaten. Fresh vegetables, aside from the readily available greens and flowers above mentioned, are rarely consumed.

Many foods of considerable importance in the diets in other parts of Mexico are infrequently used in this region, probably for economic reasons. These include most fruits and vegetables, bread and other wheat or other grain products, rice, coffee, sugar, *pastas* (various flours and pastes), lentils, peas, broad beans, and peanuts.

Corn in the form of tortillas provided 77 percent of the calories and 73 percent of the protein in the Otomi diet. These amounts are in accord with the diets shown in table 4.1 and with Sanders's assumption that 80 percent of total calories were derived from corn in the pre-Columbian diet. Pulque contributed at least 12 percent of the calories and 6 percent of total protein, exceeding the total contribution of animal protein. Pulque also contributed 10 percent of the thiamine, 24 percent of the riboflavin, 23 percent of the niacin, 48 percent of the vitamin C, and 20 percent of the iron in the Otomi diet. Truly, pulque was a staple food and a particularly significant source of vitamins.

This example was detailed at length because it illustrates the ethnocentric fallacy that a nutritious diet is not possible without large amounts of animal protein, particularly from domesticated herbivores. In particular, it reveals the potential dietary contributions of pulque and other unusual foods. It is appropriate to stress here that the Aztecs of the Basin of Mexico, with irrigation and lakes as resources, lived in an extraordinarily more productive ecological zone than the Otomi. The Aztecs were not only as omnivorous as the Otomi but drew from a much larger and more varied food base. The array of products available in a modern supermarket may seem diverse, but is really quite restricted compared to the range exploited by most hunter-gatherers. The Aztecs, hunters and gatherers until settling down in Tenochtitlan, never abandoned their earlier dietary habits; rather they added foods, such as corn, beans, squash, and chilies derived from agriculture. They did not, as some have claimed, eat weeds, rodents, and insects because they had exceeded the carrying capacity of the territory and were desperate for protein. The fact is that they simply coupled their hunter-gatherer past with their peasant agricultural present in dietary terms, just as they retained previous shamanistic elements in their now formal state religion. Hernández, who had the viewpoint of a European farmer, recoiled at the dietary habits of the Aztecs and incidentally gave evidence of the range of their larder:

> Thus these Western Indians ate tadpoles with pleasure, which our compatriots were horrified to see or even to name. They don't refuse fried locusts and ants and they consider exquisite many things which no other inhabitants of the world would eat. They not only eat tadpoles but they sell them everywhere in the markets, prepared and presented in different ways and they are not considered entirely bad nor distasteful to the palate. (Hernández 1959: vol. 2, 391)

═ Animal Protein Sources

In addition to an extensive list of fruits and vegetables consumed by the Aztecs, Sahagún enumerated more than forty varieties of edible water fowl (1950–1969: book 10, 79; book 11, 117–136). The Aztecs ate practically every living thing that walked, swam, flew, or crawled, including armadillos, pocket gophers (*tozan*), weasels (*cozatli*), rattlesnakes, mice, iguanas, and as well as domesticated turkeys and dogs. They also ate a large variety of fish, frogs, aquatic salamanders (*axolotl*), fish eggs, corixid water beetles (*axaycatl*) and their eggs (*ahuauhtli*), and dragonfly larvae, all obtained from the lakes in the Basin. Eaten land insects included several varieties of grasshoppers, ants, and worms (Sahagún 1950–1969: book 11, 58–98; Hernández: vol. 2, 390–396). Added to bounty of the Basin of Mexico per year were probably three million ducks and over one million fish. Some of the fish were especially nutritious, for example, *charales*, a small dried fish of the *Atherinidae* family. Eaten whole, it is composed of 61.8 percent protein and has three times the daily requirement of niacin per 100 g and a significant amount of vitamin A (see appendix A).

Bresani (1976) commented that small animal species can contribute substantially to human nutrition and that their meat does not differ significantly from beef and pork, (table 4.4). The table shows that protein contents are all comparable but larger species have more fat and consequently more calories. The proportions of amino acids are also similar (see appendix A), so that the meats are equivalent in quality. However, small and large animals differ in relative production efficiency. Bresani points out that smaller animals eat less, grow faster, have higher feed conversion rates, and are greater in number per unit area than larger animals, which makes them more efficient converters of plants into protein. Thus, the Aztec consumption of small animals is an advantage rather than a disadvantage because it uses the ecological system more efficiently.

Insects also are potentially an enormous source of protein because of their immense reproductive capability. For example, if all the offspring produced by one cabbage aphid during one season survived to maturity, collectively they would weigh more than the earth's entire human population (Taylor 1975: 62). In some insect species, a single pair can produce up to 47 million offspring per month and, under controlled conditions, 25 generations per year, amounting to an annual total of about 570 million. Egyptian

Table 4.4. *Nutritive Values of Meat from Various Animal Species*

	PROTEIN, %	FAT, %	ENERGY, CAL/100 g
Beef	18.7	18.2	244
Pork	17.5	13.2	194
Poultry	18.2	10.2	170
Armadillo	29.0	5.4	172
Guinea pig	19.0	1.6	96
Rabbit	20.4	8.0	159
Iguana	24.4	0.9	112
Hare	21.0	5.0	135

Source: Bresani 1976.

locusts lay up to 6,000 egg capsules on an area of one square milli-meter. Since each capsule contains 30 to 35 eggs, the number of eggs averages 195,000/mm^2 (Ramos de Elurdoy 1982: 73–74). Insects are also nutritive. Some eaten by the Aztecs and still eaten today have been analyzed and prove to be an excellent source of protein (table 4.5). The quality of insect protein, in terms of essential amino acids, is comparable to that of animals, being significantly deficient only in methionine and tryptophan (see appendix A).

By resorting to different species according to their maximum availability, the Aztecs could consume insects year round. This refutes the argument that insects are only a seasonal resource. McGregor (1984: 157) listed a possible menu schedule for insects in the Basin of Mexico:

1. *Jumiles (Atizies taxcoensis)*
   Start September, maximum November, end March/April
2. *Escamoles (Liometopum apiculatum)*
   Start April, end May
3. *Ahuauhtle* (eggs of *Corisella texcocana*)
   April–May; a bigger second crop May–July
4. White maguey worms (*Aegiale hesperiaris*)
   April–August
5. *Chilocuiles (Cossus redtenbachi)*
   July

Table 4.5. Protein Content of Mexican Insects (Dry Basis)

Axayacatl *(Corisella* ssp.)	68.7%
Ahuauhtli *(Corisella* ssp. eggs)	63.8%
Jumiles *(Atisies taxcoensis)*	70.3%
Chicatana ant *(Atta mexicana)*	58.3%
Escamoles *(Liometopum apiculatum)*	66.9%
Red *maguey* worms *(Cossus redtenbachi)*	71.0%
White *maguey* worms *(Aegiale hesperiaris)*	62.0%
Grasshoppers *(Sphenarium histrio)*	30.9%

Source: Adapted from Ramos de Elurdoy 1982: 86.

6. *Chicatana* ants *(Atta mexicana)*
   June–September
7. Grasshoppers *(Sphenarium histrio; Trimeropis* sp.)
   May–September
8. Axayacatl *(Corisella texcocana)*
   Available year round

The most famous of the invertebrates is the corixid water beetle (axayacatl, *Corisella* sp. and *Krizousacorixa femorata*). Sahagún did not say it is edible in book 11 of the *Florentine Codex*, where he described it, but he did include it in the list of foods sold in the market. Hernández (1959: vol. 2, 390) gave the earliest description of the use and abundance of this insect in the Basin of Mexico. He said it

> can be gathered with nets at certain seasons in the Mexican lake in such copious amounts that little balls made from great numbers of them mashed and mixed together are sold in the markets year round. The Indians cook them wrapped in corn husks in nitrous water and they than constitute a good food, which is abundant and not disagreeable.

Ancona (1933) confirmed that prodigious quantities of the beetle larvae are produced in March, so many that they turn the water cloudy. He reports observations of approximately 200 larvae in each cubic decimeter, equivalent to 200,000/m³. If we assume that the larvae are all within a depth of one meter, each one percent of the surface area of Lake Texcoco would produce over 2,000 tons of protein.[7] The larvae lay eggs first with the beginning of the rainy season in April and a second, bigger crop during May, June, and

July. The eggs also are eaten and were called ahuauhtli ("water amaranth") by the Aztecs. Hernández (1959: vol. 2, 392) also described the beetle eggs:

> a great quantity of a certain substance called *ahuauhtli* with a fishy taste, is taken out of the Mexican lake. It looks like a poppy seed and it is the eggs of the *axayacatl*, of which we spoke earlier. It is gathered by throwing into the lake, where the waters are most turbulent loosely twisted cables as thick as a man's arm or thigh. The [eggs] shaken and swirled adhere to these, from which the fishermen remove them and store them in large vessels. They make tortillas [from it] similar to corn ones, or the balls they call *tamales* in the native tongue, or they save it split up into portions and wrapped in corn husks in order to prepare it as food, toasting or cooking it at a later time. . . . It tastes like fish or fish eggs and they say it is not an entirely bad food.

These practices continue unchanged to this day (Ramos de Elurdoy 1982; Ancona 1933; Deevey 1957: 225). People on the shore of Lake Texcoco make bundles of reeds that they submerge and tie to stakes. After three or four days, the bundles, now covered with eggs, are recovered. The eggs are 0.5–1.0 mm long, may number $66/cm^2$ of surface area, and sometimes are laid so thickly that they form clusters (Ancona 1933). They are reported to be very tasty, with a shrimplike flavor. They can be dried and kept for several months for use as an ingredient of a traditional Christmas dish (Ancona 1933). Both axayacatl and ahuauhtli are rich in minerals and vitamins, particularly calcium, phosphorus, riboflavin, and an astounding amount of niacin 11.75 mg/100 g—see appendix A.

Agave plants, in addition to their own nutritive value, provide a growth medium for several edible Lepidoptera larvae, commonly called maguey worms. Motolinía (1971: 365) gave an early testimonial:

> on this *metl*, or maguey, towards the root, whitish worms develop, which are as thick as the quill of a wild turkey and as long as half a finger. When toasted and salted, these worms are very good to eat. I have eaten them many times on fast days when fish were not available.

Today they are consumed toasted, ground up, and mixed with salt, or fried in lard and eaten with tortillas. Ancona (1934) cited a nineteenth century comment on their delicious taste, "It provides such a tasty dish, that Parisian gourmets might prefer it to oysters from Ostend or swallows nests from China." There are two principal types of maguey worms: the white (*meocuili, Aegiale hesperia*), available during April to August, and the red (*chilocuili, Cossus redtenbachi*), harvested in July.

Other insects or insect products commonly eaten are ants (*Liometopum apiculatum* and *Atta mexicana*), stink bugs (*Attizies taxcoensis*), and the accumulated sugar water of honey ants (*Myrmecocystus melliger*). (See appendix A for the amino acids in the ants and stink bugs.)

The carrying capacity calculations and population estimates discussed in chapter 3 are necessarily simplistic and based on a diet assumed to be all corn, the crop for which data are readily available. This, however, may leave the impression that the diet was deficient in protein and in the amino acids and vitamins relatively lacking in corn. As our review shows, the Aztecs actually lived in a resource-rich environment and exploited a superb variety of foods, which even in small amounts would have remedied all the shortcomings of a corn diet. It is also clear that the Basin of Mexico was not populated near the limit of its carrying capacity and that the Aztecs were neither malnourished nor suffering from protein or vitamin deficiencies. In fact, they were probably much better fed than the modern Mexican population, which in 1973 had an average diet of 1,787 cal per capita per day and some 45 percent of which never eats meat (Ramos de Elurdoy 1982: 17–18). Since dietary adequacy affects the ability to resist infections and to recover from illnesses, it is important to establish that all Aztecs, including commoners, were well fed, as a baseline from which to evaluate their health status.

5

Epidemiology

═ **Pre-Conquest Diseases**

Scholars in this field generally agree that pre-Columbian America was relatively free of disease (Cook 1946; Newman 1976; Viesca Treviño 1984a) and suggest several explanations. One is that the migration route through the Bering Strait taken by the original settlers of America served as a "cold filter" (Stewart 1960). That is, the long passage at low temperatures served to destroy pathogens, particularly those that spent part of their life cycles outside the host's body. Insect vectors would have also been eliminated by cold. Another suggestion is that the relative scarcity of domesticated animals, compared to the Old World, diminished the number of possible zoonotic infections and parasites, such as trypanosomal diseases or malaria (McNeill 1976: 51). A third factor is the relatively late date of urbanization in Mesoamerica, which denied such "crowd diseases" as measles, smallpox, typhus, and cholera the fairly large population in close contact needed for the pathogens to maintain themselves. For example, the measles virus cannot survive below a threshold population of 300,000 to 400,000 (McNeill 1976: 60). The interval between urbanization and the Conquest was too short for these infectious diseases to develop and survive. The catastrophic decline in the population of Meso-america after the Conquest was due to crowd diseases introduced from Europe and is evidence that the natives had no immunity from previous infection. Clearly, plague, cholera, smallpox, typhus, and measles were not present in Mesoamerica before the Conquest (Viesca Treviño 1984a: 176; Joralemon 1982).

An important determinant of a population's health and ability to resist disease is its nutrition. As demonstrated in earlier chapters, the Aztecs were well nourished and would have had adequate resistance to infectious diseases. Interestingly, gout was known and fairly common, indicating the likely consumption of a large quantity of protein. This supports the contention that Mesoamerica had few famines compared to their frequency in the "famine belts" of Europe. Similarly, few epidemics were recorded in Central Mexico compared to their prevalence in Europe. It is also striking that those few epidemics were closely associated with famines caused by several years of agricultural failure (Viesca Treviño 1984a: 175), when immune resistance to infection would be at its lowest, and illnesses were secondary to physical exposure and starvation. A thorough study by Hernández (1962) listed three major and a few minor epidemics. Except for an epidemic shortly after the settling of Tenochtitlan that Hernández (1962) believed was diphtheria, the causative disease organisms are very difficult to pin down (Viesca Treviño 1984a: 176).

The identification of Aztec illnesses is controversial. Venereal diseases clearly existed in the pre-Hispanic period, associated with sexual activity and sometimes attributed to divine punishment for violations of ritual abstinence (Sahagún 1950–1969: book 1, 31; book 11, 173). Some were called "putrefaction of the member" and probably were urethritis with a mucous discharge. Jarcho's (1964) opinion was that gonorrhea did not exist in North America prior to the sixteenth century, but the possible origin of syphilis in the New World is a controversial question that has been debated since shortly after its appearance in Europe in 1494. Belief in a New World origin of syphilis, and the prevailing ideology then that God created both a disease and its remedy in the same area, led to a massive importation from the New World of two supposed cures—holy wood (*Guaiacum sanctum*) and sarsaparilla (*Smilax sp.*). These were used until replaced by mercury, an effective but dangerous medication.

Wood (1979: 230–240) summarized the arguments concerning the origin of syphilis. One view regards treponematosis (infection with spirochetes of the type that causes syphilis) as a single but extremely flexible disease that manifests itself in various ways depending on the environment and culture of the host. In this view, the manifestation in early hunters and gatherers was yaws. As people moved to cooler, drier climates and became urbanized, the spirochetes shifted their site of entry from the skin to the mucous tissue of the mouth and genitalia and became venereal syphilis. In the New World, it appeared as the tropical diseases yaws, in the lowlands, and pinta, in the drier highlands, emerging as syphilis when urban centers arose (Hudson 1965). This "unitarian theory" would make the dispute over origin moot. A second view (Hackett 1963) held that pinta was the original treponemal infection, which then underwent several mutations to yaws and syphilis, including a particularly vicious strain in Europe at the end of the sixteenth century. A third view, based on osteological evidence on both continents, squarely places the origin of syphilis in the Americas. Viesca Treviño (1984a: 182) felt that the osteological evidence is undeniable but also that syphilis was apparently fairly benign among the natives; its virulence in Europe may have been due to a lack of immunity, a mutation, or both. Baker and Armelagos (1988) concluded that a nonvenereal treponemal infection endemic in the New World was transformed into a venereal disease after its spread in Europe.

The question of the pre-Columbian presence of malaria in Mexico is also controversial. Malaria, the greatest killer of mankind, is currently a severe problem in Mesoamerica, but its existence there before the Conquest is problematical. Species of the *Anopheles* mosquito have undoubtedly always existed in the Americas, but the presence of *Plasmodium*, the malaria parasite itself, has not been established. On the one hand, intermittent fevers (tertian and quartan) were definitely mentioned in sixteenth century sources. For example, *atonahuiztli* ("aquatic fever"), associated with the Tlaloc Complex, was described as an intermittent fever with periods of chills and is cited frequently (Sahagún 1950–1969: book 11, 147, 171; Hernández 1959: vol. 1, 17, 98, 151, 152; vol. 2, 265, 394). López Austin (1980: vol. 1, 394) included an interesting passage associating atonahuiztli with an increase in the size of the spleen (*nitic mocomaltia in atonahuiztli*—literally "the aquatic fever turns into spleen inside me"). One of the effects of malaria is a greatly swollen spleen due to the accumulation of red blood cells destroyed by emergence of the merozoites (organisms formed by fission of the parasites) (Wood 1979: 253).

Proponents of a late introduction of malaria into the New World, for example, Ashburn (1947: 103–110), cite reports by early colonial explorers of thriving Indian villages and healthy individuals in tropical areas now beset by malaria. Dunn (1965) argues that *Plasmodium* is found in all the primates and mammals in Africa and Asia but only in humans and some cebid monkeys in the Americas; therefore the parasite must have been present in the New World for only a short time. Additional evidence is that many of the genetic adaptations associated with malaria in European populations, such as sickle cell anemia, thalassemia, and G-6-PD deficiency, are absent in the Americas. Wood (1975) found, by tests on human volunteers, that the female *Anopheles gambiae* mosquito, carrier of the malaria parasite, prefers to feed on humans who have O-type blood. She argued that the presence of malaria would therefore favor other blood groups in evolutionary selection. In the absence of malaria, blood groups A and B would be disfavored because they give rise to mother-child incompatibility.

Investigations of Asian populations, who have common ancestors with the populations of the New World, indeed show a comparatively high incidence and wide dispersion of A and B groups, while aboriginal populations in the New World, free from European or African contacts, all tend with few exemptions to have blood group O. This finding is inconsistent with a long duration of malaria in America, given the selective disadvantage of blood group O in the presence of *Anopheles* mosquitoes. One possible flaw in this argument is that *Anopheles gambiae*, the primary vector in Africa, is not the predominant species in the New World. There is no guarantee that other *Anopheles* species have the same preference in blood groups. Even though the Aztec atonahuiztli is provocatively similar to malaria, the evidence seems to favor a post-Conquest introduction of the disease to the Americas.

Newman (1976) listed what he considered to be aboriginal New World diseases:

1. bacillary and amoebic dysentery
2. viral influenza and pneumonia
3. various arthritides
4. various rickettsial fevers such as verruca, or Carrion's disease, insect borne and altitude localized
5. American leishmaniasis (protozoan)–espundia and uta
6. various viral fevers
7. American trypanosomiasis, such as Chagas' disease

8. round worms; especially ascarids
9. non-venereal syphilis and pinta
10. nutritional deficiency diseases such as goiter
11. bacterial pathogens such as streptococcus and staphylococcus
12. salmonella and other food poisoning agents
13. tuberculosis

Some of these, such as verruca and uta, are of South American origin and thus not applicable to Central Mexico. The existence of goiter, particularly in goitrogenic areas, is clear from ceramic figurines. Other nutritional deficiencies, as shown previously, are unlikely to have been suffered by the Aztecs. Newman's list encompassed all the diseases considered serious threats to public health by Cook (1946), with the exception of yellow fever. Viesca Treviño (1984a: 183) believed that yellow fever was found in tropical jungles, and afflicted the Maya, but not within the confines of the Aztec area.

Apart from listing which New World diseases were or were not present in the pre-Hispanic period, we must consider their relative prevalence. On the basis of specific citations by early sources and the number and types of remedies used, Viesca Treviño (1984a: 176) believed that gastrointestinal infections were the most frequent Aztec ailments. Cook attempted a quantitative comparison between European and Central Mexican disease experiences, using the following guideline:

> Now in a general way and in spite of various disturbing factors any civilization relying upon an herbalistic rationale of medicine inevitably tends to find a preponderance of medicines for those ailments which are both most common and most lethal. If, therefore, the Mexican native materia medica differed from that of contemporary Europe with respect to the number of remedies available for specific types of diseases, there would be a probability that the incidence of such types differed in parallel fashion between the two regions. (Cook 1946: 325)

He compared the remedies prescribed in John Parkinson's 1640 English herbal, *Theatrum Botanicum*, which he considered exhaustive and typical, with those listed in Sahagún (1956) and Hernández (1959). Deleting remedies aimed at producing a response (such as purgation) not explicitly linked to a particular disease,

Table 5.1. Herbal Remedies in Europe and Mexico

TYPE OF AILMENT	PARKINSON	HERNÁNDEZ	SAHAGÚN
1. Specifics for internal and animal poisons	5.8%	2.6%	0.0%
2. Animal parasites	1.8	1.2	1.2
3. Wounds, burns, fractures bruises	7.6	3.3	5.6
4. Female reproductive system	9.1	7.1	3.7
5. Organic ailments, noninfectious	38.4	36.9	18.5
6. Minor, noninfectious	3.4	3.4	4.3
7. Tumors and cysts	1.7	3.6	0.0
8. Skin diseases: bacterial mycotic, zootic	13.3	9.5	19.8
9. Infectious: inflammatory febrile, venereal	18.1	32.0	40.2
10. Unidentifiable	0.6	0.4	6.8

Source: Cook 1946: 326. Reprinted with permission from *Hispanic American Research Review*, vol. 26, no. 3. Copyright © 1946 by Duke University Press.

and dividing diseases into broad groups, he obtained the results shown in table 5.1. The general pattern is fewer organic and more infectious diseases in Mexico than in Europe. Table 5.2 shows the comparison for infectious diseases alone, also indicating a greater incidence in Mexico than in Europe.

From these data, Cook concluded that the Aztecs suffered extensively from infections of the respiratory and gastrointestinal tracts. Parenthetically, intestinal and acute respiratory infections (pneumonia and influenza) are still the leading causes of death in Mexico (Pan American Health Organization 1986: vol. 2, 156). Viesca Treviño (1984a: 177) stated that, given the prominence of parasitic colitis and diarrheas, hemorrhoids also would probably have been a significant problem. He also suggested that the tendency for gastrointestinal problems to increase in the spring was the likely source of an association of hemorrhoids with punishments by Xochipilli, the God of Spring (Sahagún 1950–1969: book 1, 31).

Viesca Treviño (1984a: 172) surmised that the Aztecs had a very high birth rate and high childhood mortality. He cited a study by Faulhaber (1965) indicating that more than ten percent of a

Table 5.2. Herbal Remedies for Infectious Diseases

TYPE OF AILMENT	PARKINSON	HERNÁNDEZ	SAHAGÚN
Respiratory: nose, throat, lungs	5.7%	8.5%	9.3%
Diarrhea, dysentery	2.6	9.0	13.0
Fevers, all types	3.4	7.7	13.0

Source: Cook 1946: 327. Reprinted with permission from *Hispanic American Research Review*, Vol. 26, No. 3. Copyright © 1946 by Duke University Press.

group of skeletons at Tlatilco represented children between the ages of four and six. Viesca Treviño also pointed out that perinatal and first-year mortalities would exceed even this high rate. He estimated an Aztec life expectancy of 37 ± 3 years, quite good compared to a calculated life expectancy of only 29 years for France at the end of the eighteenth century (Viesca Treviño 1984a: 173). Since some chronic diseases such as cancer, strokes, and cardiovascular abnormalities, particularly when linked to metabolic conditions such as atherosclerosis, become more frequent with advancing age, they are unlikely to be significant among the Aztecs simply because most died so young. The Aztec diet, low in saturated fats, and the high level of physical activity by the population would both minimize cardiovascular disease. Aztec sources do mention heart ailments, so these were known, but sixteenth-century European descriptions such as "heat or pain in the heart" are difficult to correlate with modern diagnoses. Most authors agree that cancerous tumors were infrequent among the population of Central Mexico (Coury 1969: 95; Viesca Treviño 1984a: 179).

Neurological afflictions occurred, but their incidence seemed to match that in contemporaneous Europe (Coury 1969: 87). Although no information relates to the incidence of convulsive crises such as epilepsy, they were known to the Aztecs and fit into the Aztec religious worldview. Convulsions were associated with the *cihuateteo*, deified women who had died in childbirth and would return to earth on particular calendrical dates (Viesca Treviño and de la Peña 1979). Anyone who met these beings could be possessed and afflicted with epilepsy and paralysis. Children were apparently considered more susceptible than adults (Sahagún 1950–1969: book 1, 19; 1956: vol 1, 49–50).

Evidence is clear for some pathologies that leave traces on bones. For example, various fractures were common, to judge from signs of repair and realignment frequent in skeletal remains (Viesca Treviño 1984a: 178) and sometimes accompanied by indications of osteomyelitis, periostitis, and pyogenic arthritis (Jaén and López Alonzo 1974: 162–163). Rheumatism in all its variations was also common and suffered by almost all individuals over 35, whose skeletons often showed rheumatic deformations of the spinal column (Viesca Treviño 1984a: 179). Various birth defects also are found in skeletal remains, but their frequency is difficult to determine. Micro- and macrocephaly, but so far no hydrocephaly, have been discovered (Jaén 1977: 352–354). Achondroplastic dwarfism was given much symbolic importance and associated with the supernatural. These dwarfs, together with hunchbacks and albinos, often appeared in the court of the Aztec *tlatoani*, or ruler (Sahagún 1956: vol 2, 315). Rickets has not been observed (Coury 1969: 96; Jaén 1977: 358); probably it was prevented by Aztec cultural practices such as prolonged breast-feeding, by adequate dietary calcium in tortillas made from corn cooked with alkali, and by year-round exposure to the sun due to the type of clothing and the tropical climate.

= *Public Health*

A discussion of Aztec health would be incomplete without mention of public health measures. A source of clean drinking water is essential, and the Aztecs were quite advanced in providing it. While London still drew its drinking water from the polluted Thames River as late as 1854, the Aztecs brought potable water to Tenochtitlan from springs on the mainland by means of an aqueduct built by Nezahualcoyotl between 1466 and 1478 (Bustamante 1984: 390). The aqueduct was described in 1520 by Hernán Cortés:

> Along one of the causeways to this great city run two aqueducts made of mortar. Each one is two paces wide and some six feet deep, and along one of them a stream of very good fresh water, as wide as a man's body, flows into the heart of the city and from this they all drink. The other, which is empty, is used when they wish to clean the first channel. Where the aqueducts cross the bridges, the water passes along some channels which are as wide as an ox; and so they serve the whole city.

> Canoes paddle through all the streets selling the water; they
> take it from the aqueduct by placing the canoes beneath the
> bridges where those channels are, and on top there are men
> who fill the canoes and are paid for their work. (Cortés 1971:
> 107–108)

A second aqueduct was constructed in 1499–1500 by the ruler
Ahuizotl when the first aqueduct became inadequate.

Although the Aztecs had no citywide drainage system, and
much of the waste water ended up in the lake surrounding the
city, they had a system to handle human waste by means of privies
in all public places and many private dwellings from which excrement was collected in canoes (Díaz del Castillo 1968: vol. 1, 278).
The excrement was applied as fertilizer on chinampas (Soustelle
1970: 32) or sold in the market to be used for tanning animal hides
(Clavijero 1971: 236). Urine was collected in pottery vessels to be
used later as a mordant for dyeing cloth (Vaillant 1966: 345).

The city was kept very clean. Streets in every quarter were
swept and watered daily by a thousand public employees
(Soustelle 1970: 32–33). Personal cleanliness and hygiene were
prized highly by the Aztecs, as suggested by many references in
early sources to soaps, deodorants, dentifrices, and breath sweet-
eners (de la Cruz 1964: 165, 171, 215; Sahagún 1956: vol. 3, 152, 174).
The Spanish conquerors were amazed at how often the Aztecs
bathed, in contrast to European practice. Motecuhzoma bathed
twice a day, and everyone bathed often (Ortiz de Montellano 1976).

The Tenochtitlan environment was obviously healthy for its
time, especially in comparison to European cities. Public and personal hygiene contributed to minimize the incidence and severity
of illnesses. The native population was basically healthy, due to
good nutrition, relative freedom from epidemic pathogens, and
sound public sanitation. However, dampness and pollution of the
lakes did promote dysentery, rheumatism, and respiratory infections. These general conditions should be kept in mind in the following discussion of Aztec medical beliefs.

Diagnosing and Explaining Illness

In any medical system causes dictate cures. It is useful to focus initially on etiology because etiological beliefs indicate the type of remedy people turn to. An illness thought of as divine punishment is likely to be countered by prayer and expiatory rites. Diseases caused by a sorcerer require the services of a counter-sorcerer for diagnosis and cure. A wound clearly suffered on the battlefield requires treatment to stop the blood and a protective dressing while it heals. All of these exist in Aztec medicine, sometimes simultaneously.[1] To some extent they exist in all medical systems. Even today someone en route to the hospital for surgery or treatment with antibiotics may stop in a church to pray for divine help and assiduously avoid black cats and ladders during the trip. Particular medical systems tend to emphasize one etiological concept over others and can be categorized as primarily natural (Western biomedicine, Ayurvedic medicine), magical (Azande), or religious (Christian Science, Mexican Spiritualism), but each retains elements of the other two.

Categories of Disease Origins

In this and the next chapters, for heuristic and analytical purposes, the causes and cures of diseases are divided into three categories: supernatural (religious), magical, and natural (physical). However, we must continually keep in mind that this is not the way the Aztecs regarded illness. Their view was holistic, as López Austin (1974a: 216–217) described:

> The origin of illness is complex, including and often intertwining two types of causes: those that we would call natural— excesses, accidents, deficiencies, exposure to sudden temperature changes, contagions and the like—and those caused by the intervention of non-human beings or of human beings with more than normal powers. For example, a native could think that his rheumatic problems came from the supreme will of Titlacahuan, from the punishment sent by the tlaloque for not having performed a certain rite, from direct attack by a being who inhabited a certain spring, and from prolonged chilling in cold water; the native would not consider it all as a confluence of diverse causes but as a complex.

Another useful distinction is between ultimate and proximate (or immediate) cause. Natural etiological systems focus mainly on proximate causes and the process of disease. Supernatural and magical systems are more concerned with ultimate cause, that is, why a particular disease affects a particular person. Again, a dichotomy may be more apparent than real. The god Tlaloc may be the ultimate (supernatural) cause of a disease, but he acts by sending an *ahuaque* (one of his helpers) to take possession of the afflicted individual. The ahuaque is a proximate cause who needs placation as much as Tlaloc.

= Supernatural Causation

The supernatural includes a wide variety of forces and beings, particularly in non-Western systems—deities, god helpers, their messengers, and spirits who reside in forests, lakes, and springs, as well as impersonal forces such as astrological influence. The Aztec world was full of supernatural spirits. They were thought to inhabit various natural settings—forests, rivers, lakes, springs, caves, gullies, and anthills—seen as points of contact with the underworld (López Austin 1988: vol. 1, 244–245). Ponce de León (1973: 122) mentioned the worship of clouds (inhabited by ahuaque) and winds (spirits called *eecame*). Ruiz de Alarcón (1982: 219; López Austin 1988: vol. 1, 245) describes a group of forest spirits called *ohuican chaneque* ("owners of dangerous places").

Since the Aztecs saw a universe composed of dichotomies, these spirits related to earth and water were by nature opposed to heat and the sky. They therefore sought to absorb tonalli, the animistic hot force found in people. Humans had to be cautious when approaching places such a forests, caves, and springs where chaneque dwelt and could harm them by drawing the tonalli from their bodies. Aztec sources are not very explicit or detailed in their descriptions of these animistic beings, but a wealth of information is available in modern ethnographic sources because spirit belief still widespread among Mexican Indians (López Austin 1988: vol. 1, 244–245).

Disease was also caused by deities, principally the omnipotent, unpredictable creator god Tezcatlipoca under one of his many names, Titlacahuan, "we-his-slaves" (López Austin 1970a). Devout Aztecs begged him for good fortune but also feared punishment by him for excessive pride in the ownership of goods or for breaking vows and shirking penitences. Plagues and calamities were seen as Titlacahuan's penalty for collective offenses or lack of faith (Aguirre Beltrán 1963: 44–45; Sahagún 1950–1969: book 3, 11). The smallpox epidemic that decimated the population of Tenochtitlan during the conquest of Mexico was seen as proof that their gods had turned against the Aztecs and weakened their resistance to Cortés. Incurable and contagious diseases of individuals, leprosy, mange, buboes, and dropsy also were attributed to the gods (Sahagún 1956: vol. 1, 272; 1950–1969: book 3, 11).

López Austin (1970a) felt that Titlacahuan was believed to be the ultimate cause of all ailments of supernatural origin. On occasion this function was delegated to lower gods who might in turn send messengers (such as the teyolia of people they had already killed) or other forces as the proximate causes of disease. As the commands descended the hierarchy, they became more precise and less flexible. The resulting ailment became more symbolic of the proximate causative agent. López Austin's supposition may be too simple and analytical to fit the large and confusing Aztec pantheon, but it is an interesting attempt to relate groups of specific diseases to particular deities. For example, failure to abstain from sex during festivals dedicated to Xochipilli-Macuilxochitl (gods of spring, sex, love, and dance) gave the offenders hemorrhoids or diseases of the sexual organs (Sahagún 1950–1969: book 1, 31). Violations of the prescribed fast during the feast of Atamalqualiztli every eight years resulted in ringworm and skin diseases (Sahagún 1956: vol. 1, 231; Garibay 1948). Xipe-Totec ("Our Lord the Flayed One"), a god of spring and of jewelers, afflicted people with body surface diseases such as pimples, festering sores, and a variety of eye diseases such as cataracts (Sahagún 1950–1969: book 1, 35). Xipe was associated with human skin because he was a symbol of the renewal of the "skin" of the earth in spring. Smallpox, unknown before arrival of the Spanish, was also attributed to Xipe-Totec (Sahagún 1956: vol. 1, 65). Atlatonan, a member of the Earth-Goddess Complex, both afflicted and cured lepers and people born with physical defects or suffering from sores (Durán 1971: 223).

The cihuateteo, the teyolia of women who had died in childbirth and who accompanied the sun from noon to sunset, were associated with the west and could return to earth only on tonalpohualli, the days in the ritual calendar associated with the west: 1-deer, 1-rain, 1-monkey, and 1-eagle (fig. 2.6). Men who encountered the cihuateteo on these nights (particularly at crossroads, which were places of ill omen) contracted epilepsy or palsy (Sahagún 1950–1969: book 1, 19; books 4–5, 41, 81, 107; 1956: vol. 1, 49–50, 349). The cihuateteo who descended on days 1-rain were the youngest and most dangerous. If a handsome man took sick during these days, it was said that the goddesses had taken away his beauty out of envy (Sahagún 1956: vol. 1, 349).

Death and illness caused by Tlaloc, Chalchiuhtlicue, the tlaloque, the tepictoton, and the auaque, members of the Rain God Complex, were often tied to rain, water, and cold. For example, Chalchiuhtlicue gave diseases to people near her domain, rivers and springs (Ponce de León 1973: 124–125). The overt character of afflictions due to members of the Rain God Complex could take the form of drowning, a lightning bolt, gout, dropsy, leprosy, rheumatism or paralysis (Sahagún 1956: vol. 1, 143; vol. 2, 143; vol. 3, 264–265; 1950–1969: book 1, 45; book 3, 49). Physical mechanisms as proximate causes will be discussed below in this chapter; the relationship of these diseases to cold and wet will be discussed in chapter 8.

People born on days 1-deer were predestined to die by drowning or lightning (Sahagún 1956: vol. 1, 332). Children with two cowlicks (called "human paper streamers") were destined to be sacrificed to Tlaloc during the month of Atlcaualo in a magical rite to bring rain (Sahagún 1950–1969: book 2, 42). (We noted previously that people sacrificed to a god were considered in a sense chosen by the god for the purpose.) Tonatiuh, the Sun God, summoned the teyolia of those who died in war (Sahagún 1950–1969: book 6, 144; 1956: vol. 2, 142), or who had been born on the day 4-movement (Sahagún 1956: vol. 1, 320; Serna 1953: 168–170), to accompany him from sunrise to noon. Merchants who died from whatever cause on a quasi-military expedition were similarly rewarded (Sahagún 1950–1969: book 9, 25; 1956: vol. 3, 33).

Particular animals were sometimes considered both the omen and the cause of illness, although the cause of a consequent death often was not specified. In some cases the teyolia of those who had gone to Mictlan, called *mictecah*, were sent as messengers by the ruler of the underworld. These teyolia were personified as owls, spiders, centipedes, or scorpions, all commonly regarded as bad omens (Serna 1953: 225; López Austin 1988: vol. 1, 339; Sahagún 1950–1969: books 4–5, 163). The owl, horned owl (*tecolotl*), and weasel (*cozatli*) were called *yaotequihua* ("war captains") and also considered messengers boding illness and death, they were not definitively labeled forms of teyolia (Sahagún 1950–1969: books 4–5, 165; López Austin 1969: 39). The *pinahuiztli*, which has been identified as a chafer or a spider and was often used for divination, was a powerful omen of illness if it entered a house or crossed someone's path (Sahagún 1950–1969: books 4–5, 169; López Austin 1969: 43). Tezcatlipoca sent a more direct message when a skunk, his nahualli, or alter ego, entered a house (Sahagún 1950–1969: books 4–5, 171; López Austin 1969: 47). Some aquatic animals—the *ahuizotl*, the *ateponaztli*, and the *acitli*—were omens but could also be the agents of death, perhaps with the assistance of aquatic teyolia (Sahagún 1950–1969: book 1, 31, 33, 68–70; López Austin 1970a). Garibay (1943–1946: 309; 1949) described one belief that has lasted until the present: the cry of a horned owl, *tecolotl* (said to sound like tecolo-o-o), signifies impending illness, doom, or death (Sahagún 1950–1969: books 4–5, 161). Garibay (1943–1946; 1949) links the origin of this belief to the owl's song and the onomatopoeic name:[2]

> the phrase of the owl seems to say according to the listener: "It harms someone, it harms someone" (*tecolo*, from *coloa*—to bend, to twist; a primitive form of the reduplicated *cocoloa*, to be sick; *cocolia*, to hate; *cocoleti*, to become thin etc. All of these have the sense of harm either physical as in *cocoliztli*—illness or *cocolia*—hate). (Garibay 1943–1946: 309)

A person's tonalli was an omen established at birth, a particular tendency and temperament that determined the person's future. Astrological dates had consequences for health. If a woman born on 7-flower fell from grace, Xochiquetzal would afflict her with piles and infections (Sahagún 1950–1969: books 4–5, 7). Men born on 1-flower who acquired wealth, ignored the gods, became proud, and failed to respect their parents risked many ailments, including blindness, decaying of the genitals, paralysis, gout, hemorrhoids, and sores (Sahagún 1950–1969: books 4–5, 24). People born on 1-monkey had to be careful outdoors on their birthday because the cihuateteo could descend on that day. Any illness contracted on that day, even if not epilepsy, would be incurable, and the doctors would give up (Sahagún 1950–1969: books 4–5, 81). Men born on 1-deer were especially susceptible to death by drowning or lightning (Sahagún 1950–1969: books 4–5, 9). The *Codex Vaticanus 3738* (1964: fol. 31, 47, 49), a less reliable source than Sahagún, listed the following birthdate omens: 7-eagle–sickly, incurable hearts; 1-wind—if people became ill with fistula or *dolor de costado* (probably pneumonia) they would not recover (Figueroa Marroquín 1957: 24); 1-snake—a limp or mild injury to a limb meant eventual loss of the limb.

Both Europeans and Aztecs had astrological beliefs but with significant differences. European astrology, originating in either Egypt or Babylonia and more fully developed by the Greeks, held that all matter was composed of four elements (air, earth, fire, and water) with four qualities (hot, cold, wet, and dry). Thus, Venus was hot and wet, Mars was hot and dry. Humans were composed of four humors (blood, phlegm, black bile, and yellow bile) also with the four qualities. Astral bodies could influence human bodies because of this commonality, their influence depending on their current position on the zodiac, an imaginary strip surrounding the heavens and divided into twelve "houses" named after the constellations they contained. An individual's lifespan and fate were determined by the zodiacal sign, called the ascendant, at the moment of birth (Castiglioni 1947: 236–242). The planets could cause diseases, and particular planets affected particular organs. For example, according to one scheme Mars was related to the gallbladder, Venus to the kidneys, and the moon to the brain, hence the term lunatic (Hartmann 1973: 150).

In Aztec astrology, the equilibrium of the universe affected the human body. The tonalli given an individual at birth depended on the date on the tonalpohualli. Unlike the European astrological beliefs, however, tonalli was a link to the gods and the heavens collectively, not to any particular planet or star. Cosmic influence for the Aztecs was much more general, and organs of the human body were not associated with individual heavenly bodies or day names in the calendar. Light rays from Venus could cause illness, but one that was poorly defined and did not affect a specific organ (Sahagún 1950–1969: book 7, 12, 62).

Some have suggested that the twenty day names in the tonalpohualli were thought to affect specific parts of the body (Aguirre Beltrán 1963: 50; López Austin 1988: vol. 1, 349) on the basis of an illustration (fig 6.1) in the post-Conquest *Codex Vaticanus 3738* (1964: fol. 73) showing the day signs attached to parts of the body.[3] The *Codex* was painted between 1566 and 1589, is annotated in Italian, and reflects the Europeanizing tendency in the Early Colonial Period (Robertson 1959: 111). Concepts in the

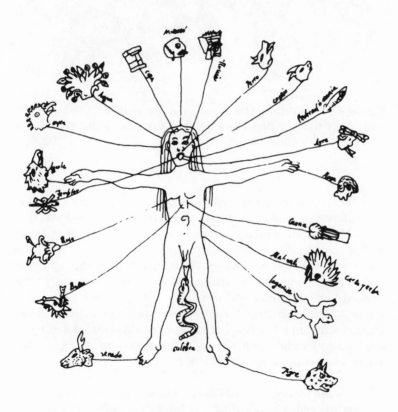

Figure 6.1.   A human figure with day signs attached to various parts of the body. This has been used to claim that the Aztecs assigned astrological signs to illness in certain organs. (*Codex Vaticanus Latinus 3738* 1964: plate 73)

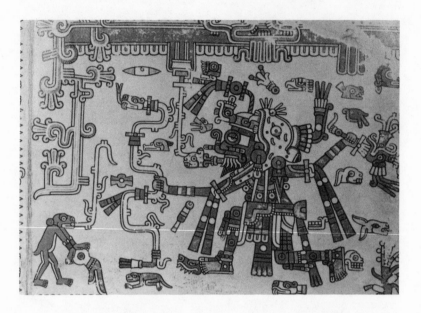

Figure 6.2. The God Tlaloc with day signs distributed over his body. Note that the distribution differs from that in figures 6.1, 6.3, and 6.4. (*Codex Laud* 1964: fol. 2)

illustration are not supported in Sahagún or other more reliable texts. However, Europeans of that period commonly believed that signs of the zodiac influenced particular organs and that astral bodies were linked to human physiology (Hartmann 1973: 132–160; Dick 1946). Contemporary European illustrations showing zodiac-body connections (Singer 1957: plate 16) could easily have swayed both the artist and the commentator of the *Codex Vaticanus 3738* to misinterpret a native Aztec tradition of superimposing day signs on a drawing of the figure of a god (figs. 6.2, 6.3 and 6.4). The concluding sentences of the text accompanying the *Vaticanus* plate support this implication:

> thus also the doctors used this figure when they cured; according to the day and the hour in which one got sick, they saw if the illness was in accord with the sign that was ruling. By which it can be seen that these people were not as savage as others have made out, since they had so much mathematics and order in their things, and they used the same means that the astrologers and doctors use among us. (*Codex Vaticanus 3738* 1964: fol. 54)

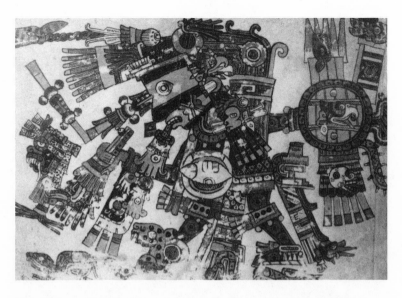

Figure 6.3.   The God Tezcatlipoca with day signs distributed over his body. The distribution differs from that in figures 6.1, 6.2, and 6.4, probably due to artistic license rather than astrological significance. (*Codex Borgia* 1963: folio 17)

A similar European distortion can be seen in the description of the Aztec calendar in Serna (1953: 117). Here the Aztec year-bearers, the only days that could initiate the year, were forced into an Aristotelian mold: *calli* ("house") was said to correspond to earth, *tochtli* ("rabbit") to air, *acatl* ("reed") to water, and *tecpatl* "flint" to fire. No evidence exists elsewhere for this type of analogy. If, in fact, the commentator of *Vaticanus 3738* were correct, the day sign associated with a given organ should vary little in different sources. This is decidedly not the case. Pre-Conquest illustrations in figure 6.2 (*Codex Laud*: fol. 2), figures 6.3 and 6.4 (*Codex Borgia*: fols. 17 and 73, respectively), and the post-Conquest *Codex Fejervary-Mayer* (1964: fol. 44—illustration not included here) show no consistency in the allocation of day signs to anatomical features either within a single codex or between codices. Furthermore, the degree of European influence in the post-Conquest *Vaticanus* is revealed by the distinct differences between the pictorial conventions in figure 6.1 (*Vaticanus*) and figure 6.3 (pre-Conquest, and presumably more authentic, *Borgia*). At present, the most reasonable conclusion is that the arrangement of day signs on the

Figure 6.4. The Gods Quetzaloatl and Mictlantecuhtli with day signs over their bodies. This distribution differs from that in figure 6.3 although both come from the same source. (*Codex Borgia* 1963: fol. 73.)

human body was a matter of individual artistic choice rather than deep astrological significance. An interesting and relevant ethnocentric comment was Sahagún's simultaneous criticism of Aztec astrology and defense of European astrology in an appendix to book 4 of the *Florentine Codex*:

> It [the Aztec calendar] is a difficult count, full of error, and with no foundation of natural astrology. For the art of judiciary astrology, common among us, is founded upon natural astrology, which is in the signs and planets of the heavens and in their courses and aspects. But this art of soothsaying followeth, or is founded upon, some characters and numbers in which no natural foundation existeth, but [are] only an artifice made by the Devil himself. Nor it is possible that any man could have made or invented this art. For it hath no foundation in any science nor any natural order. (Sahagún 1950–1969: books 4–5, 145)

Another supernatural feature of Aztec medical beliefs was the role of "bad" and "good" smells in health and disease. During a siege of Culhuacan, an episode in Aztec history, the Mexica threw worms, fish, ducks, and frogs into a fire and allowed the smoke to waft into the city. The smoke supposedly caused pregnant women to abort, people's limbs to swell, and even death from unsatisfied desires (Durán 1967: vol. 2, 93; Dibble 1963: 41–42). Although unsatisfied desires by themselves could cause illness, as when children requested but were denied pulque (Sahagún 1950–1969: books 4–5, 193), other evidence existed for the baneful effects of bad smells. The myth of the struggle between Quetzalcoatl and Tezcatlipoca for control of Tula related that the Toltecs killed a sorcerer, Cuexcoch, whose corpse emitted a strong stench that killed many of the common people (Sahagún 1950–1969: book 3, 27–28). Similarly, the hair of people who smelled a skunk and then spat would turn white, a belief linked to the mythical struggle of the gods since the skunk was a nahualli of Tezcatlipoca. Also, the smell of the *teccizuacalxochitl* plant caused the nose to swell (Sahagún 1950–1969: book 11, 209). The relationship between bad smells, disease, and the supernatural had a parallel among the Maya:

> entrance to the underworld was a cavern in Alta Verapaz. From this entrance or hole came a stench of rotting corpses and clotted blood. It was the duty of the Maya medicine man to send back through this hole the diseases inflicted by the underworld rulers on the race of men . . . Cizin [ruler of the underworld], a name derived from *cizin*, meaning "flatulence or breaking wind" attesting to the Maya view of his realm as a world of putrescence and corruption. (Coe 1975: 89)

López Austin (1988: vol. 1, 358) proposed an Aztec belief that aromatic substances repelled underworld and aquatic beings, primarily by their ability to attract tonalli, which belongs to the heavens. As with other opposing pairs in Aztec cosmology, good things—good smells in this case—are associated with the upper world, and bad things—the fetid—belong to the underworld. Hernández described the use of *apitzalpatli* to strengthen tired rulers whose duties require extra tonalli (1959: vol. 1, 3) *huacalxochitl* flowers as "highly esteemed by the Indians" and offered to *tlatoani* (1959: vol. 1, 389); and the *coatzontecoxochitl* (*Stanhopea tigrina*) as having an extremely good smell and used to prepare food for rulers who had stayed too long out in the sun (1959: vol. 1, 119). The *Badianus Codex* listed prescriptions that support the etiological significance of smells. One of the components of a complex regimen to "cure the fatigue of rulers and public officials" is good-smelling flowers (de la Cruz 1964: 193). Fragrant flowers were similarly prescribed as a cure for melancholia (de la Cruz 1964: 195). Epilepsy was treated with a variety of magical remedies including the "good" smells of burning mouse nests and copal incense, presumably to drive away spirits that possessed the victim.[4]

= Magical Causation

The magical is not easily separated from the supernatural. However, if a sorcerer causes an illness, or if a disease-inducing action can be analyzed in terms of two well-known laws of magic, the law of contact and the law of similarity, then we can regard the cause as primarily magical. According to the law of contact, objects once in contact with an individual can continue to exert influence at a distance, and can be used magically. According to the law of similarity, actions performed on one plane can act analogously on another. A classic application of the law of similarity is sticking pins into a voodoo doll to cause corresponding pain in the person whom the doll represents. In some cases we have categorized illnesses as magical simply because they are not obviously either religious or natural.

López Austin (1967) described several kinds of Aztec magicians and sorcerers. The most common generic name for sorcerers was *tlacatecolotl* ("owl man"). As a symbol of the underworld, the tecolotl was an appropriate namesake. The tlacatecolotl derived his powers from being born on designated days of the tonalpohualli, 1-wind or 1-rain (Sahagún 1950–1969: book 10, 32). Sorcerers found their birthdays and days numbered nine especially propitious for casting spells and doing their evil deeds (Sahagún 1950–1969: books 4–5, 102). If captured, sorcerers could be destroyed by cutting the hair from the crown of their heads, which led to a loss of tonalli (their powers) and eventually death (Sahagún 1950–1969: books 4–5, 102). Often the spells cast by sorcerers took a physical form. Hostile wishes were buried and became ants (Sahagún 1950–1969: book 4–5, 173); ailments could be left at the side of the road to be picked up by the feet of the first passerby (López Austin 1971b), causing an inflammation of feet and ankles called *xoxalli*. Although xoxalli is also described as a "cold" disease, due to the overheating of tonalli or the presence of twins, it is clearly related etymologically to sorcery since the root *xoxa* means "to bewitch someone" (López Austin 1988: vol. 1, 261–262).

Belief in magical illnesses due to bodily intrusion by objects or forces is widely dispersed in the world (Sigerist 1967: 128). Tlatlatetecolo threw spells, which could lodge in various parts of the body and "manifest" themselves as pieces of bone and obsidian, removable only by another sorcerer (Serna 1953: 100, 102, 106; Garibay 1973: 151). A unique type of sorcerer could drive people insane by harming their teyolia, the *teyollocuani* by "eating" their hearts and the *teyolpachoani* by "squeezing" their hearts (Ruiz de Alarcón 1982: 65; Sahagún 1950–1969: book 11, 3). Similarly, sorcerers called *tecotzcuani* ("he who eats the calves of people") caused a type of muscular ailments (Garibay 1943–1946: 168).

When sorcerers were not specifically mentioned in the sources, the line between magic and divine etiology becomes especially blurred. Urinating or sitting on certain plants and flowers, or sometimes just smelling them, could cause disease. Urinating on *huerembereque* risked severe itching and skin eruptions (Hernández 1959: vol. 1, 399); urinating on a cacao bush (*Theobroma cacao*) or its flowers could result in a skin infection called *xixiotiz* (Garibay 1973); and urinating on *aquiztli* (*Paullinia fuscescens*) could lead to blisters all over the body (Sahagún 1950–1969: book 11, 131). Imitative magic was involved since aquiztli was also used to cure blisters. Hernández (1959: vol. 1, 38) tells of the belief that people who slept in the aquiztli's shadow would lose their hair. The best known examples of the phenomenon of imitative magic involve the *omixochitl* (*Polyanthes tuberosa*) and *cuetlaxochitl* (*Euphorbia pulcherrima*). Omixochitl flowers were said to resemble the human penis and, if urinated or stepped on, would cause the genitals to decay (Sahagún 1950–1969: books 4–5, 183). Others (Garibay 1973: 143: Serna 1953: 218; Hernández 1945: 128) told of the Aztec belief that smelling or urinating on the flowers produces *xochicihuiztli* ("hemorrhoids"). Sitting on, stepping over, or smelling cuetlaxochitl would cause infection of women's sexual organs (Sahagún 1950–1969: book 4–5, 183; Serna 1953; 218; Hernández 1945: 128). Another aspect of the genital effect of these plants was their relation to the deities Xochiquetzal and Macuilxochitl, who were gods of singing, art, love, and spring and associated with diseases of the genitals and the anus. Macuilxochitl used xochicihuiztli as a punishment for violations of ritual (López Austin 1969: 189).

Examples of contact magic are scarcer. Often they involved love magic, for example, in the belief that a woman might ensure her husband's love and forestall his cruelty to her by putting menstrual blood in his food or drink (Garibay 1967: 45). A sorcerer called a *tetlepanquetzqui* ("he that prepares fire for people") also used this type of magic. He would adorn a rod with funerary paper and offer ritual mourning food to it for four days, the usual period of mourning. The following day, an intended victim would be invited to partake of the food while the sorcerer secretly wished the victim an imminent death. More directly, the sorcerer would acquire some of the prospective victim's hair and burn it to represent incineration of the soon-to-be cadaver (Garibay 1943–1946: 169; 1967: 21–21; López Austin 1967).

For a culture that continually made symbolic connections between objects in the natural world and the supernatural, belief in sympathetic magic came easily. People who ate the seeds of *pochotl* (*Ceiba* sp.), a large bulky tree, would fatten like the tree to an incapacitating degree (Hernández 1959: vol. 1, 300). The milky latex of *cuetlaxochitl* leaves were used to increase the milk produced by wet nurses (Hernández 1959: vol. 1, 319). If children licked the *metate* (grinding stone), their teeth would fall out (Sahagún 1950–1969: books 4–5, 188). A pregnant woman who ate the tamales that stuck to the pot would have trouble giving birth (Sahagún 1950–1969: books 4–5, 188).

This type of magic pertained particularly to pregnant women because a mother's actions often sympathetically affected the fetus. Pregnant women were not supposed to chew gum or tar because their babies would be born with swollen lips or a cleft palate and not be able to suckle. If corn cobs were burned, the baby would be born with pockmarks. A mother-to-be should not see someone hanged because the baby would be born with the umbilical cord around its neck (Sahagún 1950–1969: books 4–5, 188–190). Other prohibitions on pregnant women included not to look at anything red or the child would be born crosswise; not to sleep during the day or the child would be born with abnormally long eyelids; and not to get overheated in the steam bath or the baby would swell from the heat (Sahagún 1950–1969: book 6, 156).

One Aztec magical belief about lunar eclipses, still widely held in some areas, was that a pregnant woman who goes out at night and sees an eclipse of the moon will give birth to a baby with a harelip. Although *tochicihuiztli* ("the mark of the rabbit") one of the Aztec names for harelip (López Austin 1988: vol. 2, 182), seems to refer to the rabbitlike appearance, the reference is actually more complicated. The name derives from the Aztec creation myth, which tells how the face of a rabbit was imprinted on the moon. The name for a lunar eclipse *metzqualoniliztli* ("the eating of the moon"), attributes the diminishing crescent to successive bites. By sympathetic magic, a fetus exposed to the moon in eclipse would have a bite taken from its lips—*tenqualo* ("the eating of the lips"), yet another name for harelip. To protect her baby from this fate, a pregnant woman could place an obsidian knife on her belly (Sahagún 1950–1969: books 4–5, 189; Serna 1953: 214; Garibay 1973: 145). Hundreds of years later in Tecospa, an Aztec village near Mexico City, Madsen (1960: 9) found this belief practically unchanged except that the knife was made of metal because obsidian was no longer available.

The flesh of the hummingbird, a medicine against pustules, was said also to cause sterility. The hummingbird was a symbol of the solar god Huitzilopochtli ("Hummingbird on the Left"), patron god of the Mexica. Since Nanahuatzin ("the little pimply one") had become the sun, according to the creation myth, the therapeutic use of the solar god's symbol had both religious and magical connotations. Divine power was invoked (religious) to get rid of pimples like the sun's (magical law of similarity).

Interestingly the hummingbird, with its strong ties to Aztec religion, has been used in Mexico for centuries as a magical amulet combined with prayer to seduce the opposite sex. The desiccated bird, which resembles a penis and acts by imitative magic, is wrapped in silk and placed in a bag for carrying. Women carry female hummingbirds while men carry male ones. The men's prayer, supposed to be said every Friday before an image of Christ, is a revealing example of syncretism:

Oh divine hummingbird! You who give and remove nectar from the flowers; you who give life and teach women to love. I take refuge in you and your powerful emanations so that you will protect me and give me the ability to make love to any woman I want whether maid, married, or widowed.

I swear by all the Sacred Apostles not to cease for a moment to adore you in your sacred reliquary so that you will give me what I ask, my beautiful hummingbird. (Quezada 1975b: 106)

Diagnosis of Supernatural and Magical Ailments

The diagnosis of ailments of supernatural origin (divine punishment) required the services of specialists to identify both the offended deity and the appropriate propitiatory rites. Propitiation in advance might also achieve prevention (Ruiz de Alarcón 1982: 208). Diagnosis and prognosis of illness could be done by a nahualli (Garibay 1943–1946: 167). Such specialists were also called *paini* ("one who drinks medicine"), because they ingested hallucinogens to go to other worlds to consult with supernatural beings, returning with a diagnosis, a prognosis, and, if possible, a remedy (Ruiz de Alarcón 1982: 87–88). This procedure is characteristic of shamanism.

Martínez Cortés (1965: 99) described two types of diagnoses associated with supernatural ailments. One dealt with the pathologic future of a healthy individual and usually consisted of omens associated with the calendar animals, insects, or sexual conduct. The second type dealt with the course of an existing illness and often involved magical procedures and/or the use of a paini. The diagnosis of most illnesses of divine origin has been described in previous chapters. What follows primarily concerns diseases of magical origin.

According to Aztec mythology, the creator gods Tezcatlipoca and Huitzilopochtli gave the tools and knowledge for illness divination by magical means to the primordial couple Oxomoco and Cipactonal (*Historia de los Mexicanos* 1973: 25; *Codex Borbonicus* 1974: fol. 21–22; Sahagún 1950–1969: books 4–5, 4). Figure 6.5 shows the couple using the techniques of divination. One of the procedures shown in the *Codex Borbonicus* was that of the *mecatlapouhque* ("the one who counts [the meaning] of the ropes"). The diviner tied several knots along a cord in the presence of the patient and then pulled strongly on the ends of the cord (López Austin 1967: 106). If the knots untied or loosened, the patient would recover; if the knots tightened further, the patient would die (Torquemada 1975–1983: vol. 3, 131; Garibay 1943–1946: 241).

A second procedure shown in the *Codex Borbonicus* was the casting of corn kernels, *tlaolxiniani* (fig. 6.6). Different sources offer slightly different versions of the procedure. According to Sahagún's informants, the kernels were tossed onto a cloth; if they fell in orderly rows or piles, the patient would do well, but if they scattered, death was near (Garibay 1943–1946: 240). According to

Figure 6.5. The couple Oxomoco and Cipactonal who the Aztecs believed originated the art of magical illness divination. (*Codex Borbonicus* 1974: plates 21 and 22)

the *Codex Magliabechiano* (1983: fol. 77v), twenty kernels were cast upon a cotton cloth placed upon the ground. The patient was destined to die soon if the kernels formed a circle, but would recover if they fell in parallel lines so that a line could be drawn through the middle without touching a kernel. If one kernel fell on top of another, the illness was due to homosexual activity. More elaborate descriptions were given elsewhere (Ruiz de Alarcón 1982: 213–215; Serna 1953: 264–265; Gómez de Orozco 1945: 45). Practioners shaped the corn correctly for use in divination by biting off the points of well-shaped kernels. Nineteen or twenty-five were chosen, twelve or nineteen were arranged in a pattern on a white cloth, and seven or nine, respectively, were actually tossed in the air to land on the cloth. If the kernels fell with the bitten end down, the patient would recover; otherwise the method would not cure.

Incantations normally were an integral part of divinations, to judge from Ruiz de Alarcón. Other sources that consistently minimized the magical aspects of Aztec medicine omitted mention of

Figure 6.6.   Divination of illness by casting corn on mats. Resulting patterns indicated cause and prognosis. (*Codex Magliabechiano* 1983: plate 78)

the incantations. The language in these incantations, *nahuallatolli* ("nahualli speech") was esoteric, complex, and full of elaborate metaphors. The following is typical:

> Come (H), Tlazohpilli, Seven Snake [i.e., the maize kernels]. Come, Five-tonals-owners, One-courtyard-owners [i.e., the hands]. It is the instant at last [i.e., now is the time to act]. Let us go in order to see their joke that is his worry. Will it perhaps be by-and-by tomorrow? It will indeed be immediately at this instant. It is I in person. I am Cipactonal. I am Old-Man. Soon I will look inside my book, my mirror [i.e., the thrown magic kernels], [to see] if a medicine will match him [i.e., be appropriate to him] or if he will get worse. (Ruiz de Alarcón 1984: 154)

In figure 6.7 the kernels are shown being cast into a bowl of water (López Austin 1967: 105; Garibay 1943–1946: 240; Ponce de León 1973: 132; *Codex Magliabechiano* 1983: fol. 76v). Ruiz de Alarcón (1982: 217) stated only that a number of kernels were thrown into a fairly deep vessel of water; Ponce de León said the number was seven. A good omen was sinking of the kernels to the bottom. Prognosis was poor if the kernels floated or submerged but remained suspended.

Figure 6.7.   Divination of illness by casting corn in a bowl of water. (*Codex Magliabechiano* 1983: plate 77)

Water was also used in other ways. A child was placed over a bowl of water to determine if it had lost tonalli. Loss of tonalli was confirmed if the child's face cast a shadow, while a shiny reflection indicated a healthy condition. To cure loss of tonalli, the *tetonalti-que* ("tonalli providers") resorted to incantations invoking water and tobacco and then frightened the child by spraying the magical water on its face (Ruiz de Alarcón 1982: 223–229). Other cures for loss of tonalli are given in chapter 8. Serna (1953: 98–99) described a procedure involving the color and behavior of the water in a bowl placed at the head of a sick person. If the water turned red, the illness was due to sorcery; if it turned yellow, the illness was divine in origin; and if it spun around and diminished, the patient was soon to die.

Illness was also divined by measuring the number of the diviner's handspans that fit on the left forearm of the patient (López Austin 1967: 103). The measurement was repeated several times from the elbow to the tip of the patient's hand. The diviner would prepare by first rolling dried tobacco between his hands[5] and intoning an incantation in which the hand was referred to as "Five-tonalli," the forearm of the patient as "Our Jade Ladder," and the diviner as "the Lord of Mictlan" (Ruiz de Alarcón 1982: 202–209). The patient was expected to die soon if the diviner's handspans and the patient's forearm matched perfectly; if the diviner's last handspan went slightly past the patient's hand tip, death would result but might be delayed; if most of the diviner's hand was left over, recovery was possible (Serna 1953: 258–259). The procedure was also used to determine if the illness was natural or due to enmity or sorcery.

The Aztecs were accurate observers of nature and capable of precise clinical diagnosis and prognosis (Fernández del Castillo 1942). The *ticitl* ("doctor") was able to control the process of divination when he could foresee the outcome of a presenting condition. Ruiz de Alarcón (1982: 224) pointed out that this was indeed possible. Fresh plump corn sinks in water; old or wornout kernels float. The proper kind was chosen to ensure an obvious prognosis. Similarly, divination of tonalli loss in children could be rigged. The child to be tested was placed with the sun behind to cast a shadow or in front to avoid a shadow (Ruiz de Alarcón 1982: 224). A skilled practitioner was also allowed to cast a desired pattern of corn kernels or to tie slip knots as required (Serna 1953: 265).

Manipulating the odds does not mean that the healers were cynics perpetrating fraud. Shamans often hide small objects or threads in their mouths, which they then spit out (often covered with blood produced by biting their cheeks) and claim to have sucked out of a patient. Levi-Strauss (1967: 167) said that workable magic requires three complementary aspects: The sorcerer must believe in the efficacy of the technique; the patient must believe in the sorcerer's power; and the group's expectations must provide a kind of gravitational field defining the relationship between the patient and the healer. Thus, Quesalid, a Kwakiutl shaman, who believed that object extraction was a trick, nevertheless found that he could cure people and was more effective than other shamans who did not produce "evidence" of the disease (Levi-Strauss 1967: 169–170). Although the shaman's manipulation primarily enhances the faith of the patient and confirms the group's expectations, the experience of successful healing also induces self-confidence in the shaman.

Two other animal omens not mentioned previously were

used. A *pinahuiztli* (chafer) was captured, spat upon (law of contact), placed in the center of a quadrant drawn on the ground, and observed. If the insect moved first toward the north, the direction of Mictlan, land of the dead, the omen was of death (Sahagún 1950–1969: books 4–5, 169–170; López Austin 1969: 42–45, 183). Similarly, if a sacrificed decapitated quail headed north when placed on the ground, the owner of the house was destined to sicken and die (Sahagún 1950–1969: book 9, 37–38; López Austin 1969: 156–157).

= Natural Causation

In some cases of natural etiology the connection between cause and effect is obvious. The Aztecs generally treated bites by poisonous snakes, spiders, and insects by cutting into the bite, sucking out the poison, and rubbing tobacco into the wound (Sahagún 1950–1969: book 11, 77, 88–94; Hernández 1959: vol. 2, 386–388). Until the recent development of snake serum, conventional medical treatment has not been markedly superior to that of the Aztecs. The Aztecs believed that sprains, dislocations, and fractures damaged the blood, producing swelling, inflammation, and possible infection at the site of the trauma. If left untreated, the disease lead to swelling of the belly, constant coughing, and finally heart damage (López Austin 1988: vol. 1, 167–168). Treatment, of questionable benefit, involved drawing blood to reduce swelling and medicines to stimulate blood circulation and prevent stagnation (Sahagún 1950–1969: book 10, 161–162). However, the Aztec treatment of wounds, to be discussed in chapter 7, was effective even by modern standards.

Etiologic beliefs can be deduced, even when not explicitly stated, from the kinds of remedies prescribed for particular ailments. Confidence is greater if different sources support and supplement each other, especially when the existence of corroboration in more reliable sources like Sahagún makes data from sources such as Hernández and de la Cruz more credible. One example is headaches, apparently believed to derive from an excess of blood in the head. (Interestingly, some migraine headaches are now known to arise from constriction of blood vessels, which inhibits the flow of blood around the cranium and produces pressure and pain). The rational cure for an excess, according to the Aztecs, was to remove some blood from the head. Since very few orifices allow repeated and controlled blood removal from the head, an induced nosebleed would seem optimal. The following describes an Aztec treatment for headache:

> Its cure is to inhale an herb named *ecuxo* [sneeze plant], or to inhale [green] small tobacco . . . and if it becomes worse, an herb named *zozoyatic*, which is dried and pulverized, is to be inhaled. . . . If it has worsened, if the nose [treatment] no longer helps, the use of an obsidian point, of incising, of bleeding there [on the head] is necessary. (Sahagún 1950–1959: book 10, 140)

Violent sneezing was intended to make the nose bleed, as indicated by the plants used: *ecuchoton* ("sneeze plant"), whose name is self-explanatory, tobacco (snuff) and *zozoyatic* (*Schoeno-caulon coulterii*), whose contents severely irritate the mucous membrane. The first of Sahagún's versions is even more specific, "It is cured by bleeding our nose. . . . It is cured with sneezes, sniffing tobacco in order to sneeze. . . . [It is cured] by cutting with obsidian" (López Austin 1972: 141, 145). When a nosebleed failed to bring relief, the head was pierced directly with a flint, a serpent fang, a feather quill, or thorns (Hernández 1959: vol. 1, 121, 123, 360, 376). Sahagún's descriptions are echoed in other early colonial sources, including Hernández and the *Badianus Codex*. Hernández cited twenty-nine headache remedies that included picietl and zozoyatic. All were smoked or otherwise inhaled and are often described as stornutatory (Ortiz de Montellano 1979b). The data support the hypothesis that the Aztecs attributed headaches to excessive blood in the head, and the agreement among descriptions in different sources buttresses our confidence that the information is valid and that the judicious use of sources such a Hernández is warranted.

The Aztecs also regarded changes in the animistic forces of tonalli, ihiyotl, and teyolia as primarily natural in their effects on health. Although the deities imbued humans with these forces in the first place, the forces thereafter acted with little or no supernatural intervention. So behaviors or circumstances that diminished their strength in a person could be considered natural, proximate causes of illness. Treatment addressed the proximate causes.

Sexual behavior in particular was felt to have a strong influence on animistic forces and hence on physical health. In fact, the threat of illness from loss of these forces was actually much more effective than Aztec legal sanctions as a deterrent against adultery and aberrant sexual practices. The close connection between physical illness and sexual lapses comes through in the word *cocoxqui,* which means not only sick or weak but also homosexual or effeminate (Molina 1970: fol. 24r). Thus, a unique aspect of Aztec physiology was that the beliefs about animistic forces served as a primary method of social control, augmented by laws prohibiting adultery and unnatural sexual practices. Another aspect is that, despite the natural character of animistic forces, the Aztecs treated illnesses related to ihiyotl with primarily magical rather than the empirical cures that would be expected for a natural ailment. The proximate cause of diseases associated with teyolia was natural, but its ultimate cause might be supernatural possession.

The physiological connotations of the animistic forces had concrete expression. A person who led a moral life was said to have a unified liver (*cemelli*); immoral sexual conduct altered the liver and caused its involuntary emission of ihiyotl. The transgressions of an adulterous woman would cause birth difficulties even if she used the oxytocic remedy derived from opossum tail. The magical cure was to introduce the woman's saliva into her vagina to induce the birth (Serna 1953: 250). Homosexuality and other proscribed sexual activities also could have physical effects leading to the release of ihiyotl in an emanation that might cause economic reverses such as frozen crops, the death of livestock, or the failure of business deals (Ruiz de Alarcón 1982: 182), and affect weak persons such as babies or old men. Young turkeys, notoriously fragile, would die and fall flat on their backs at the arrival of an adulterer. Any falling and landing on one's back was said to be a sign that one's spouse was an adulterer (Sahagún 1950–1969: books 4–5, 191; 1956: vol. 2, 36). Mice were apparently hardier than turkeys; an adulterer's ihiyotl could not kill them but would make them eat holes in baskets. Mice were used as diagnostic indications: they would gnaw holes in a husband's mantle if his wife was an adulteress and in a wife's skirt if her husband was guilty (Sahagún 1950–1969: books 4–5, 191–192; 1956: vol. 2, 35; Garibay 1967: 52).

Diseases collectively called *tlazolmiquiliztli* ("filth death"), caused by the ihiyotl of sexual sinners, can be divided into three types: an illness of children who wake up screaming and sick, or certain incurable adult diseases; similar infantile diseases involving fainting or epileptic fits and called *tlazolmimiquiliztli* ("frequent filth death"), with the reduplicative element *mimi* perhaps referring to the repetitive nature of the seizures; and *netepalhuiliztli* ("harm due to dependence on another"), a weight loss and wasting away as if from consumption (Ruiz de Alarcón 1982: 192; Serna 1953: 270–272). The emanations of adulterers could also give their spouses *chahuacocoliztli* ("illness due to adultery") (Karttunen 1985: 45). The nature of these diseases and children's special susceptibility to them were easily joined to the European concept of the "evil eye" (chap. 8).

Released ihiyotl could also do harm in other ways. Nanahualtin were able to emit ihiyotl at will to afflict people directly, or indirectly by lodging the harm first in animals, which then later affected people (López Austin 1988: vol. 1, 371). Although not clearly designated as ihiyotl, a force emanated from the left arm or left middle finger of cihuateteo, women who died during their first childbirth. Such a finger placed in a warrior's shield made him fearless and enabled him to capture many enemies (Sahagún 1956: vol. 2, 180). Sorcerers called *temacpalitotique* ("those who make people dance in the palm of their hand") could render householders unconscious by knocking a cihuateteo's left arm against the door of the house to facilitate robberies (Sahagún 1950–1969: books 4–5, 103–105; 1956: vol. 2, 180; López Austin 1966).

Several groups of ailments were associated with deficiencies in the tonalli essential for proper growth, vitality, and strength. One group was related to sexual intercourse. We have already referred to the Aztec belief that loss of tonalli due to too early an initiation or too much sex prevented young men from reaching full development. This was particularly crucial for nobles who needed extra tonalli in order to help them rule. It was also peculiarly a male problem because sex did not exhaust women's reserves. Excessive sex could be fatal to men: infusion of an extract from the serpents *mazacoatl* and *tlalmazacoatl* killed by inducing continuous, uncontrollable erections and ejaculations (Sahagún 1950–1969: book 11, 80; Hernández 1959: vol. 2, 373). Sex-related illness could be contracted in various ways according to two general principles. On the one hand, sex required energy and also diminished tonalli, so one should be strong before engaging in it. On the other hand, retained semen, lack of orgasm, or interruption of coitus also were harmful. Here, as in other cases, health resulted from avoiding extremes. Having sex was dangerous while recovering from an illness or too soon after giving birth; too little tonalli was available. Also, both parties would suffer, and the weakened one would relapse (Sahagún 1950–1969: book 11, 155, 176, 183). Hernández (1959: vol. 2, 70) lists *netlapilpatli* as a remedy for the fever caused by having sex with a recently delivered woman.

Tonalli could be diminished, thus precipitating illness, not only by excessive or ill-timed sex, but by the participants being scared while having sex. Tonalli left the body during sex, as it did in sleep, and a sudden interruption prevented its full return to the body. Symptoms of such ailments were loss of weight and impotence (Sahagún 1950–1969: book 11, 145, 174, 183; book 6, 117). Erotic and wet dreams, *cochitemictli* ("sleep that produces illness"), produced semen and thus also diminished tonalli (López Austin 1988: vol. 1, 295; Garibay 1967: 22, 54). However, the Aztecs seem to have believed that something was also introduced into body and became corrupted, as evidenced by remedies such as *iztac cuahuitl* (Sahagún 1950–1969: book 11, 148) used to cure the disorder by releasing blood and a purulent fluid. This disorder could also afflict women; the prescription then included the plant *coanenepilli* added to the iztac cuahuitl (Sahagún 1950–1969: book 11, 148). Hernández (1959: vol. 1, 198, 431) says that iztac cuahuitl and coanenepilli were also capable of releasing retained semen. Failure to achieve orgasm in coitus and semen retention during erotic dreaming apparently were thought to produce decayed matter that must be expelled to affect a cure (Sahagún 1950–1969: book 1, 145, 183).

Physiologically, the first few months of development of a fetus was thought to require periodic coitus because semen, *xinachtli* ("seed" as in Latin and English), provided nourishment. This belief is incorporated in the word for "son of a whore," *tlanechicolpiltontli*, literally "son of things that are piled together; son of a cooperative mixture" (López Austin 1988: vol. 1, 298). However, coitus was forbidden during the last months of a pregnancy because the semen, no longer needed, would merely decay. The baby would then be born covered with a sticky white film that would adhere to the womb and make birth difficult (Sahagún 1950–1969: book 6, 78–82). López Austin (1969: 202) remarks that this describes an infection of the amniotic sac. Naeve (1979) shows that intercourse during the final month of a pregnancy is more likely than abstinence to cause infection and perinatal mortality.

Tonalli could leave the body not only during sleep and sex, but on various other occasions such as severe fright. Molina (1970: fol. 71r) translates the word *netonalcaualtiliztli* as "he who is frightened of something," but it literally means "the loss of tonalli" and has survived to the present in the form of the folk illness *susto* ("fright"). Children were considered more susceptible to the loss of tonalli because their skull sutures were not fully closed, and one symptom of a child's loss of tonalli was a sunken fontanelle (López

Austin 1988: vol. 1, 206). Except for shamans, tonalli could leave the body only for short periods of time because it was essential to life. While outside the body, tonalli was in constant danger of attack by *chaneques*, cold wet beings that lurked in woods and caves and desired tonalli's heat and power (López Austin 1988: vol. 1, 226). Free tonalli therefore usually sought protective cover: a bowl of water, a small animal, or, best of all, the plant *tlacopatli* (*Aristolochia mexicana*) (López Austin 1988: vol. 1, 219). Hernández (1959: vol. 2, 130–131) notes that necklaces of tlacopatli were used medicinally, particularly to treat all kinds of cold diseases. Young men who were pledged to enroll in a school were draped with tlacopatli necklaces, which were left at the school as a token that they would return when they were old enough (Sahagún 1950–1969: book 3, 63; 1956: vol. 1, 305) because, as stated in an earlier Sahagún manuscript, "supposedly his tonalli was in his necklace; supposedly his tonalli remained in it" (López Austin 1971: 134–135). Tlacopatli necklaces and roots were variously used to treat sick children (López Austin 1971b: 134–135). For example, the root was placed on a sick child's crown, the hair of the crown was raised, and the healer beseeched sun to restore the child's lost fate and health. After the incantation, a line was drawn with tobacco from the tip of the nose up to the crown of the head (Ponce de León 1973: 125, 131–132) so that the "hot" substance would seal shut a possible avenue for further loss of tonalli.

Hair had power because of its proximity to tonalli and could protect against its loss. Endangered people, especially relatives left behind by *pochteca*—long-distance merchants—setting out on a journey, would not cut or wash their hair until the danger was over (Garibay 1967: 10–11; Sahagún 1956: vol. 1, 246, 345). Some of the *piochtli*, hair on the back of children's heads, was never cut, to prevent loss of tonalli (Torquemada 1975–1983: vol. 3, 131).

As indicated before, the relationships between tonalli and temperature were complex. Tonalli was hot but also regulated body temperature; loss of tonalli produced a fever rather than a chill (López Austin 1988: vol. 1, 216–217). Walking or working heated the tonalli; resting cooled it. The verbs *ceceltia* and *cehuia* meant either "to rest" or "to cool down a hot thing," and *tonalcehuia*, "to rest one who walks," was literally "to cool down tonalli." However, as tonalli heated up, other parts of the body, particularly the feet and the navel, cooled down. Much walking produced "cold feet," a disease called *xoxalli*. The remedy was to rub the feet with tobacco (*picietl*, *Nicotiana rustica*) and *axin*, fat from the scale insect *Llaveia axin* (Jenkins 1964). Picietl was classified as "hot" because it was used against "cold" diseases such as gout and swelling of the belly (treated by placing picietl in the navel), and as a rubdown to relieve fatigue (Sahagún 1950–1969: book 11, 146). It was also recommended as a stomach rub against physical cold (Alva Ixtlilxochitl 1975: vol. 2, 74).[6] Axin was "very hot" and used for gout and to rub the feet of walkers to protect them from the cold (Sahagún 1950–1969: book 10, 90). If tonalli was cooled too rapidly, its strength was diminished. López Austin (1988: vol. 1, 216) cited a phrase meaning "to be frightened" that literally says "my tonalli leaves me as if someone had poured water over me." If one of the temacpalitotique sorcerers sat down while fleeing from a crime, he allowed his tonalli to cool and would become too weak to rise (López Austin 1966).

= *Mixed Causation—Cold and Phlegm*

Cold and phlegm diseases had natural proximate causes but ultimate origins that might be supernatural. The Aztec holistic approach to etiology often makes it impossible to classify diseases into neat Western categories. However, it is clear from the concept of tonalli and its attributes permeating the Aztec worldview that disease involved an imbalance of hot and cold. We will discuss a few of the many references to heat, cold, and disease that were not tied to tonalli.

The *Badianus Codex* mentioned illnesses caused by excessive heat in the head, the heart, and the eyes (de la Cruz 1964: 153, 159, 161, 179, 199). Only cold in the abdomen is explicitly named (de la Cruz 1964: 183). The recommended cures were a mixture of the rational, for example, herbs and precautions such as avoiding bright sunlight and sweat baths, and the magical, such as the use of gem stones and parts of animals (Viesca Treviño and de la Peña 1974). Hernández (1959: vol. 1, 120) mentioned heat in the stomach due to exposure to the sun. Dental caries were attributed to extreme variations in temperature. Very warm food was not recommended in any case; to avoid caries, you waited a while between consuming warm things and having a cold drink (Sahagún 1950–1969: book 10, 146). This principle explains the prohibition against chewing green stalks of maize at night in order to avoid caries. The stalks must have been classified as very cold, since the problem could be avoided by heating the stalks in a fire before chewing them (Sahagún 1950–1969: books 4–5, 194). It was also safe to chew them during the day because the heat of the sun would counter-vail against their innate coolness. Finally, a particular form of diar-rhea, called *aminaliztli* in Molina (1970: fol. 5v) and *apitzaliztli* in Sahagún, was caused by drinking cold water after eating raw vege-tables (Sahagún 1950–1969: book 11, 209).

That gout, *coacihuiztli* (literally "the stiffness of the serpent"), was a "cold" disease can be inferred from a variety of data indicating a mixed etiology. It was sent by deities of the Tlaloc Complex as punishment (the ultimate cause) but was also said to derive from cold winds and winds issuing from caves (the proximate cause). The disease was thus linked to the cold and wet typically associated with aquatic deities. Also, the recommended remedies reveal il-luminating similarities and interconnections. For example, picietl and axin, which we have seen were clearly classified as "hot," were applied locally for gout (Sahagún 1950–1969: book 10, 90; book 11, 146). Another group of locally applied remedies for gout is interest-ing. All hallucinogenic if ingested, they included *toloa, tlapatl, mixitl* (varieties of *Datura stramonium*), *ololiuhqui* (*Turbina corym-bosa*), and *nanacatl* (*Psylocybe* sp.) (Sahagún 1950–1969: book 11, 129, 130, 147). The strength and quantity of tobacco ingested by na-tives would have made it also a hallucinogenic. These plants were used in two senses: rationally, as "hot" remedies (by analogy with axin and picietl) opposed to a "cold" disease and in a religious mys-tical sense, as entheogens (god-containing) remedies whose deities revealed themselves upon ingestion. The plants were thus partly a means of evoking influence over the ultimate divine cause of the disease, Tlaloc and his helpers.

Combining mutually reinforcing information from various sources, as demonstrated in determining the Aztec etiology of headaches, together with etymological analysis can help to define a group of diseases thought to be caused by phlegm. Some of these ailments are tied to the Tlaloc Complex as an ultimate cause and are also discussed in chapter 8. López Austin (1974a: 220) sum-marized their effects:

> Accumulation of phlegms, white, green, and yellow, which pen-etrate the nerves, head, and chest [sic]. These phlegms produce pressure on the heart, fever, pulsations of the temple, and mus-cular and nervous trembling. The pressure on the heart pro-ceeds from one side, injures the viscera (the organs of thought), and produces a loss of consciousness.

This loss of consciousness could also be manifested in a form re-sembling epilepsy when the cihuateteo were encountered upon one of their earthly visitations:

> Perhaps one of them would have twisted lips, or a shrunken mouth; or be cross-eyed or of weak vision; or his lips would quiver or quake; or he would be maddened—a devil would pos-sess him and he would foam at the mouth; or he would have withered, twisted arms; or be lame, etc. (Sahagún 1950–1969: book 4–5, 81)

We have described the functions of the animistic force teyolia, which resides primarily in the heart. An accumulation of phlegm in the chest affected the heart and could produce madness, stu-pidity (since the heart was the seat of the intellect), fainting, or epi-lepsy. Definitions of terms in Molina's dictionary show some of the range of effects:

yollomimiqui (literally "dead heart")—dumb, stupid
teyocuepaliztli ("turning around of the heart")—rave, crazy
yolaactiuechtiliztli ("plummeting of the heart")—tired soul
neyoltequipacholiztli ("crushing of the heart")—tired soul
yolpatzmiquiliztli ("crushing of the heart")—epilepsy
yolzotlaualiztli ("fainting of the heart")—epilepsy.

Some of the synonyms for madness in the *Florentine Codex*—*teyollomalacacho* ("heart spins around"), *teyolpatzmiqui* ("crushed heart"), *poliuiznequi yiollo* ("heart wants to die")—also support this interpretation (Sahagún 1950–1969: book 11, 129, 130, 155). A direct connection with phlegm was shown by another illness de-scribed as *alaoac quipoloa toiollo* ("phlegm destroys the heart"), treated by *haacxoyatic* (*Ipomoea capillaceae*) (Sahagún 1950–1969: book 11, 181). The treatment of mental stupor in the *Badianus Codex* involved the herb *tlatlacotic* (*Bidens* sp.), to induce vomiting, followed by *yolloxochitl* (*Talauma mexicana*, "heart flower"), to "ex-pel violently the bad humor in his chest" (de la Cruz 1964: 213), both consistent with a phlegm-based etiology and accompanied by distinctly magical elements. Also, Hernández (1959: vol. 1, 185) described the *tlahuelilocaquahuitl* ("tree of madness") as a cure for heart and chest ailments. An empirical remedy for these ailments, such as a purgative, diuretic, emetic, or diaphoretic, would be in-tended to remove the offending phlegms from the body. In fact, prescribed remedies are said to have just such effects, which supports the etiological hypothesis.

The basic names for fever were *totonqui,* derived from tona, and *motlehuia* ("to have fire within"). The Florentine Codex describes several localized fevers, for example, fever in the head, the stomach, and the eyes (Sahagún 1950–1969: book 11, 165, 176, 154). However, these seemed to be surface manifestations of fever located inside the body. Three further distinctions can be made— *iztac totonqui* ("white fever"), *matlaltotonqui* ("green fever"), and *atonahuiztli* ("aquatic fever"). Iztac totonqui is also usually described as a "high fever" and associated with swollen body tissues (Sahagún 1950–1969: book 11, 163, 168, 177). Matlaltotonqui was reputed to surface on the body as dark spots and bruises (Sahagún 1950–1969: book 11, 158, 181). In both cases references were made to removing phlegm from the body. If this etiology was indeed correct, and if the Aztecs were good observers of nature, one would expect fever remedies to in fact produce diaphoretic, emetic, diuretic, or purgative effects. An analysis of all the botanically identifiable fever remedies in the *Florentine Codex* has shown that seventy percent contained chemicals known to do what Aztec etiology required; that is, they were emically effective (Ortiz de Montellano 1979b).

"Aquatic fever" differed from the other two types but was still related to phlegm. It was also associated with intermittent recurrent chills closely resembling malaria. Its relation to the Tlaloc Complex is evident in the "water" found in its name and in its treatment with the herbs *iztauhyatl* and *yauhtli,* both closely identified with Tlaloc. The aquatic connection is reinforced by the presence of the Aztec hieroglyph for water in the illustration accompanying the disease *tlanatonahuiztli* ("aquatic fever of the teeth") in the *Florentine Codex* (Sahagún 1950–1969: book 10, 146). Atonahuiztli resembled gout in the application of the hallucinogenic remedies nanacatl, toloa, ololiuhqui, and *peyotl (Lophophora williamsii),* except that they were ingested rather than topical. As in the case of gout, the treatments had several objectives. As emetics, they were eaten or drunk to eliminate phlegm, the proximate cause. Symbolically, they were "hot" substances counteracting a "cold" disease. Finally, as hallucinogens, they were a pathway to the gods, the ultimate cause. The divine connection of atonahuiztli can be deduced from the fact that Sahagún eliminated these types of illnesses from later versions of his work, although they appeared in the *Primeros Memoriales.* These ailments were *tlallatonahuiztli,* "aquatic fever of the earth," and *yoallatonahuiztli,* "aquatic fever of the night." Other eliminated diseases were also clearly related to religion—*necihuaquetzaliztli,* "the elevation of a woman," which referred to women who died at first birth, and *netlahuitequiztli,* "lightning strike," which referred to death ordained by Tlaloc.

Phlegm-caused diseases can be analyzed further. The ones mentioned so far entailed a gradual accumulation of phlegm. However, it seems that a sudden terrifying experience also was deemed able to precipitate a harmful accumulation of phlegm, which would explain illnesses such as a "near miss by a lightning bolt," or "fear from a strong wind." Both were related to the Tlaloc Complex, since Tlaloc dispatched people to Tlalocan through death by lightning. Similarly, the fright of a sudden encounter with the cihuateteo at a crossroads might generate sufficient phlegm to precipitate an attack of epilepsy. A survey of several sources shows considerable overlap in the remedies described for fear, wind and lightning, and epilepsy. Sahagún, Hernández, the *Badianus Codex,* and some other early Colonial sources cited the same remedies for the same illnesses (Ortiz de Montellano 1979b), remedies also used for other phlegm diseases and sometimes mentioned in connection with the removal of humors. For example, Hernández (1959: vol. 2, 10) said that:

> [*quauhyayahual*] evacuates bile and mucus by the upper or lower channel. . . . Applied locally to the head, it helps the frenetic and crazy; removes fever; strengthens the head of newborn babies; and returns to consciousness those wounded by lightning, apoplectics, and others who have lost consciousness for any reason.

Tlatlanquaye (Iresine calea) was used to treat epilepsy and fever and against the heart being squeezed (Sahagún 1950–1969: book 11, 174). Table 6.1 summarizes the network of relationships between phlegm diseases and remedies in various sources. A wide consensus is evident.

López Austin (1988: vol. 1, 232, 340, 355) surmised that epilepsy, lightning, fear, and madness were due to possession by either the cihuateteo (in the case of epilepsy) or the *ahuaque* and *ehecatotontin,* the teyolia of those chosen to die by Tlaloc. Some of his evidence was: 1) Molina (1970: fol. 122r) defined *tlayouallotl mopan momana* as "fainting or illness that covers the heart" and

Table 6.1. *Phlegm Illnesses and Remedies in Various Early Souces*

REMEDIES	EPILEPSY	WIND & LIGHTNING	HEART, FEAR	FEVER	HUMORS
Aacxoyatic		B	F	F,H	F,H
Copal		B		F	F,H
Ecapatli	F,H			H	H
Iztauhyatl	B,F	B,F	B,F	F	F
Maquauhpatli		B,F,H			
Quauhyayahual		B,F,H	H	H	H
Quauhyyauhtli	H	B,H			H
Tlatlacotic		B,F	B	F,H	F
Tlatlanquaye	F,H	B,H		F	H
Youalxochitl	F,H		H	F,H	H
Yyauhtli		B,F,H	F,H	F,H	F

Source: Adapted from Ortiz de Montellano 1979b.
B = *Badianus Codex* (de la Cruz 1940), F = *Florentine Codex* (Sahagún 1950–1969),
H = Hernández (1959).

tlayouallotl tepam momama as "epilepsy or similar ailment," but both expressions contain the root *tlayoalli* implying that shadows have invaded (López Austin 1988: vol. 1, 232). 2) *Aacqui* is defined as "struck by lightning or crazy" but literally means "he who has suffered an intrusion" (Molina 1970: fol. 1r; López Austin 1988: vol. 1, 355). 3) In naming the epilepsy produced by encountering the cihuateteo, the *Florentine Codex* always used the expression *ipan oquizque cihuapipiltin*, "the women came upon him" (Sahagún 1950–1969: book 1, 19; book 4–5, 81). 4) Those affected by lightning and struck dumb were described by the phrase *in iuhqui itech quinoa*, "who is as if possessed" (Sahagún 1950–1969: book 11, 188). 5) Hernández (1959: vol. 1, 185) described the *tlahuelilocaquahuitl*, "tree of madness or rage," as both good for the heart and famous among the natives for driving out evil spirits, two views that were not incompatible. Possession by supernatural spirits would have been the ultimate cause that generated phlegm, the proximate cause.

Aztec etiology was too complex and holistic to be encompassed by the proposed categories. Our attempt, for heuristic reasons, to distinguish separate religious, magical, and natural causes ran into difficulties in cases such as phlegm diseases where all three modes were present. It might be better to think of Aztec etiology as a continuum between natural and supernatural realms, with particular ailments variously located between the extremes, rather than as a system of distinct and exclusive entities. Fevers caused by phlegm were primarily natural in their believed origin, while lightning-induced phlegm diseases were of mixed natural and supernatural origin. In either case, however, cures incorporated rational and effective remedies (either by emic or etic standards) as well as magical and religious elements.

Chapter

7

Curing
Illness

Even though modern biomedicine is the dominant ideology in the West today, even middle-class people still resort to a wide variety of unorthodox healing procedures (McGuire 1988), many with a supernatural (either magical or religious) basis. One of the reasons given for such types of healing in this scientific age is that biomedicine is not holistic and does not adequately take into account the interaction of mind and body. Scientific biomedicine is not able to deal satisfactorily with questions of meaning, such as "Why me?" and "Why do the good and/or the innocent suffer?"

Since different healing techniques are considered to influence different spheres, their simultaneous application need not cause a conflict. Thus, someone suffering from rheumatoid arthritis may dutifully take high doses of aspirin and follow other prescriptions but also wear a copper bracelet (a magical approach) and attend Pentecostal services religiously on the theory that being "born again" and in harmony with God will enlist Jesus in a cure. The feeling and appearance of symptom relief is the crucial determinant. As devotees of faith healing proclaim, healing results from becoming "right with God" or "in harmony with the universe." The consequent joy and inner peace constitute a "cure" regardless of objective medical assessment (McGuire 1988: 35–36, 68–70). The

Aztecs also were interested in the relief of illness, and whether relief was primarily psychological/symbolic, due to a placebo effect, achieved by the empirical use of remedies with active ingredients, the result of surgery, placation of spirits, or magic, or a combination of all the above did not matter to the healer or the patient. To us it is interesting and valid to make such distinctions about Aztec remedies and to determine the effectiveness of their cures in both their terms and ours. This chapter will assess such effectiveness in both terms.

═ *Religious Cures*

Diseases attributed to deities were treated with rituals, offerings, confession, expiation, and prayer. People afflicted by Xipe-Totec with skin and eye diseases, for example, would vow to take part in the spring celebration dedicated to him during the month of Tlacaxipehualiztli (Sahagún 1956: vol. 1, 65). They would volunteer to wear the flayed skins of men who had impersonated the god (according to the Aztecs the men had *become* the god) and had been sacrificed at the beginning of the celebration (Durán 1971:172–185). By wearing the skins for several days, they symbolically participated in the renewal of the "skin" of the earth. They might also assist in disposing of the god-impersonators' skins at the end of the festival month (Shagún 1950–1969: book 2, 57–60; 1956: vol. 1, 148–149).

The Aztecs believed that hills and volcanoes were inhabited by Tepictoton, assistants to and members of the Tlaloc Complex. Hills were filled with water and were the source of rain. Some hills were inhabited by members of the Rain-Agriculture-Fertility Complex, such as the tzentzon totochtin. People afflicted with ailments attributed to cold and the Tlaloc complex, such as coacihuiztli (gout), paralysis, and cramps, and those who feared death by drowning would vow to take certain actions in the month of Tepeilhuitl, "Feast of the Hills" (Sahagún 1950–1969: book 1, 47–49; 1956: vol. 1, 72–73). At that time they would fashion anthropomorphic representations of some of the volcanoes and mountains from amaranth dough. These representations had faces with black beans for eyes and were decorated with *amatl* paper spotted with rubber. The images would be honored for four nights with music, copal incense,[1] and other offerings. On the fifth day, the images were decapitated, and the amaranth dough was taken to the temple. Decapitation mirrored the sacrifice and subsequent decapitation of several female "god impersonators," killed to obtain agricultural fertility (Sahagún 1950–1969: book 2, 131–133). The ceremony was supposed to cure participants of coacihuiztli. Durán (1971:255–256, 453) wrote that the models of hills were augmented by crooked and gnarled tree branches fashioned by the priests into mock snakes by covering them with amaranth dough and adding eyes and a mouth. After pretending to kill the snakes, the priests distributed them among the lame and those suffering from pustules and paralysis. Consumption of this dough was called "god eating" and had curative properties. Motolinía (1971: 53) also described this practice.

Confession played an important role in therapy. Mendieta (1971: 281) pointed out, perhaps hyperbolically, that doctors treated mild illnesses with herbs but attributed chronic or severe illnesses to sin for which confession was the principal medicine. For example, the cure for illness due to sexual immorality was confession of sin before the goddess Tlazolteotl, who in this guise was called Tlaelquani ("filth eater") because she would eat the sins and thus cleanse the sinner. After confession, penances such as fasting or puncturing the tongue with small reeds were imposed (Sahagún 1956: vol. 1, 51–53). Confessions were also made before Tezcatlipoca (Sahagún 1950–1969: book 6, 29–31) who was said to wash the sinner clean and prevent future illness. Durán (1971: 245) mentioned that people would go to the river in honor of Xochiquetzal to cleanse their sins. If they did not do so, they would be susceptible to contagious diseases such as leprosy sent by the gods as punishment for sins. Similarly, a bath accompanied by incantations, copal smoke, and fire was said to be the only cure for tlazolmiquiliztli (the "filth" disease caused by ihiyotl). The bath was called *tetlazolaltiloni*, "the means of washing someone with regard to filth," that is, sexual misconduct (Ruiz de Alarcón 1984: 136).

Confession and suggestion are extremely important psychological therapies in non-Western cultures and are also used in Western psychotherapy, although not emphasized in theory (Torrey 1986: 88). Confessions share guilt, provide emotional catharsis, replay painful experiences (abreaction), and can be therapeutic by themselves. As Balzer (1983) indicated, confession imparts a sense of participation during therapy sessions, and painful experiences reorganized through abreaction and transference can be rationalized and explained by a familiar mythical system. Suggestion can take many forms (Torrey 1986: 90), ranging from a direct command to cleansing with water to ceremonies of death and rebirth, as well as sacrifice. Examples have already been given and others will appear in discussions of mythical cures.

Specific deities can be tied to cures in several ways. Some diseases have specific etiologies involving particular deities to be implored for healing, for example, diseases associated with the Tlaloc Complex (listed in chap. 8) and diseases of the sexual organs that fall in the province of Macuilxochitl (Sahagún 1956: vol. 1, 58–59). Gods might be identified with particular medicines, such as the "black water" used to cure children in Ixtlilton's temple (Sahagún 1956: vol. 1, 62), the "divine medicine" kept in Tezcatlipoca's temple (Durán 1971: 115–118), and the salve, *axin*, associated with the goddess Tzapotlatenan (Sahagún 1950–1969: book 1, 17). Also, a group of goddesses belonging to the Earth-Mother Complex—Tozi, Ixcuina (the goddess of birth), and Tlazolteotl—were patrons of the health professions. These goddesses had dual characteristics of fertility and birth as well as illness and death so that their association with both midwives and doctors was logical. Tozi was the patron of diviners, of people who extracted objects from others, and of the steam bath often used medicinally (Sahagún 1950–1969: book 1, 15–16; 1956: vol. 1, 47–48). The goddesses were also patrons of the neighborhoods where midwives and doctors lived. Ochpaniztli, the feast month dedicated to Tozi, was primarily oriented toward agricultural fertility and harvest, but doctors and midwives participated prominently in the festivities (Sahagún 1950–1969: book 2, 118–126; 1956: vol. 1, 190–196).

Magical Cures

Symbolic healing and the placebo effect are intimately involved with magical and religious cures, and language is an important component. The Aztecs had a high regard for language and a strong concern for its proper usage. They distinguished between fine rhetorical speech, such as huehuetlatolli and poetry, and the speech of commoners, or even worse, of foreigners. The esoteric and complex nahuallatolli was used by sorcerers and in curing rituals such as those described in Ruiz de Alarcón. The Aztec view of language, different from our own, is somewhat analogous to that of the Navajo.

According to Witherspoon (1977: 9), Navajo philosophy regards mental and physical phenomena as inseparable; thought
and speech have creative power that can affect the world of matter
and energy. Thus, the Navajo gods (*Diyin Dine'e*) created the universe by their thought, expressed in speech and song. This differs
from "In the beginning there was the word." We interpret the
phrase metaphorically, whereas the Navajo literally attribute creative power to the words themselves (Witherspoon 1977: 17).
Words are also power in another sense. The Navajo, as were the
Aztecs, are animistic: the Diyin Dine'e are the inner forms of
various natural phenomena. A particular Dinin Dine'e, good or
evil, can be controlled by knowing its symbols, especially its name
(Witherspoon 1977: 61, 84).

This is an example of what Torrey (1986: 18) called the Principle of Rumpelstiltskin, the first component of all psychotherapy,
both Western and non-Western: the very act of naming the problem is therapeutic. The name of the principle derives from the
nineteenth century tale by the Grimm brothers in which a magical
dwarf threatens to take away the queen's baby unless she guesses
a man's name correctly. After appropriate suspense, she learns the
name—Rumpelstiltskin—and lives happily ever after. According
to Torrey (1986: 18–70), the principle works (in any system of psychotherapy) only if the therapist shares some of the patient's
worldview of the disorder being treated. The curer must use symbols appropriate to the worldview of the patient, a sand painting
for the Navajo, suctional extraction of small objects by a shaman,
heart-bypass surgery for a middle-class American male. In the
Navajo realm of medicine, curing rituals, often called "sings," re-
enact the myth of creation of the world through songs, prayers,
and symbols such as sand paintings. The patient thereby enters
mythical space and identifies with the power of the deities to be
restored to harmony with the world (Witherspoon 1977: 25). Language plays an essential role, much different than in the West:

> *In the Navajo view of the world, language is not a mirror of real-
> ity; reality is a mirror of language.* The language of Navajo ritual
> is performative, not descriptive. Ritual language does not de-
> scribe how things are; it determines how they will be. Ritual
> language is not impotent; it is powerful. It commands, compels,
> organizes, transforms, and restores. It disperses evil, reverses
> disorder, neutralizes pain, overcomes fear, eliminates illness,
> relieves anxiety, and restores order, health, and well-being
> (Witherspoon 1977:34).

Viesca Treviño (1984b: 207) suggested that Aztec incantations were directed to the supernatural to discover the origin of illness and where the cure is to be found. Incantations were forms of logotherapy, cure by words. As Eliade (1963: 29) stated, such medical chants make the myth of the origin of the medicine an integral part of the cosmogonic myth and thus symbolize the transition between the imaginary and the real. The Aztec use of the nahuallatolli tied the world of myth to the act of curing and gave the healer additional symbolic powers. As in the case of the Navajo, the Nahuatl words identified the curer with the supernatural, and the use of names, often obscure, summoned the help of good deities or banished the influence of evil ones.

> I am the priest,
> I am the lord of transformations.
> I brought you, venerable pearly head [the finger tip].
> Seek the green palsy.
> Seek it, venerable pearly head. (López Austin 1970b: 7)

> Well now! Please come
> One Serpent, Two Serpent,
> Three Serpent, Four Serpent.
> Why do you harm the magic mirror,
> the magical eye?
> Go somewhere.
> Go away somewhere.
> If you do not obey me, I will call
> she of the jade skirt,
> she of the jade blouse,
> because she will drive you away,
> she will scatter you,
> she will swiftly scatter you upon the plain. (López Austin
>   1970b: 7)

Torrey (1986: 11) argued that the techniques of Western psychiatrists are, with few exceptions, on exactly the same scientific plane as the techniques used by shamans. A review of over one hundred other studies of the effectiveness of different psychotherapeutic techniques found no one significantly superior statistically to any other (Luborsky, Singer, and Luborsky 1975). The common factors in success were precisely those identified by Torrey (1986:18–69) as essential in both psychotherapy and non-Western healing: 1) a shared worldview, which enables the Principle of Rumpelstiltskin to work; 2) the healing effects of a benign human relationship between the healer and the patient; 3) client expectation of cure, which enlists the healing power of faith by means such as catharsis and abreaction (relief of a suppressed emotion by talking about it); and 4) a sense of mastery over life's adversities acquired through the learning instilled by the therapy. All of these were applied by the Aztec.

Moerman (1979; 1981; 1983) discussed symbolic healing and the placebo effect in several papers. Infectious disease is multifactorial, caused as much by a patient's lack of immunity as by invasion by bacteria and viruses. The immune system is also involved in both neoplastic and autoimmune chronic diseases. Immunological resistance is influenced by a wide range of psychological and social factors. The hypothalamus is the most probable link between the mind and the immune system. Moerman listed three components in the healing of any sickness that can function simultaneously. First, healing may be autonomous, since many conditions are self-limiting and resolved by the body's immune system. Second, specific medical treatment may cure the illness. The third healing is a certain combination of the first two components, in which the *act* of medication, but *not* the content of the medication, somehow triggers the body to achieve autonomous healing (Moerman 1981). Moerman (1979) concluded that roughly half the effectiveness of modern internal medicine is due to specific medical treatment, that is, the action of the medicine itself, and the remaining half to various placebo effects, which he called general medical treatment. The placebos attain potency as symbols interpreted by the human mind.

One of the chief characteristics by which anthropologists distinguish humans from nonhumans is the ability to use symbols, particularly in the form of language. This symbolic facility developed with the evolution of the human brain in size and complexity and is a unique activity of the forebrain, composed of the thalamus and the hypothalamus. The hypothalamus performs various functions. As a neural center, it regulates body temperature, digestion, and the intake of food and water. As an endocrine center, it manufactures some hormones and regulates the levels of others in the blood. It does so through its control of the pituitary gland, which, in turn, regulates other glands such as the pineal, thyroid, gonads, pancreas, and adrenals (Moerman 1979).

When physicians or other healers prescribe inert medications (by design or otherwise), about a third of all patients feel improvement, and the physicians also detect improvement in approximately the same percentage of patients, but the two groups may not be the same patients (Moerman 1981). Because apparent therapeutic results may be unrelated to the composition of medication, new drugs, to qualify for possible acceptance, must be proven more effective than a placebo in controlled double-blind trials. Substantial placebo effects have been demonstrated in medical circumstances as diverse as pain, hypertension, wound healing, depression, anxiety, rheumatoid arthritis, warts, and acne (Moerman 1981). The effects may be mediated through the body's release of endorphins in the brain, which are analgesic and mood lifting, or through actions of the immune system. To be effective and powerful as a placebo, a symbol must be culturally appropriate and represent an aspect of the therapeutic alliance between patient and healer (Spiro 1986: 217).

Placebos are of two general types: inert, for example, the traditional sugar pill, and what Weil (1983:210) calls "active." Active placebos have physiological effects, but the effects are not related to the condition being treated. They are more common (and more effective) than inert placebos because patients sense that something is really happening to them, and their conviction can in itself initiate a therapeutic response. Weil (1983: 216, 225) suggested that the strong faith in technology prevalent in our society predisposes patients to respond better to relatively traumatic procedures than to gentler treatments, and that bitter and expensive pills also fit the patient's preconceptions of what drugs should be like. Spiro (1986: 39–40) called these "impure placebos":

> If the pill has a side effect such as the dry mouth produced by anticholinergic drugs, the placebo effect may be enhanced, the patient surmising that if the drug is strong enough to dry his mouth, it must be doing some good elsewhere. From such considerations, I assume, came all the bitter tonics and pretty pills of the old pharmacopoeia.

Weil (1983: 225) amplified this concept:

> Psychoactive drugs are model active placebos because they really do make people feel different, and because the feelings they give are neutral and ambivalent, lending themselves to the creative interpretation and the shaping of individual users.

Many Aztec treatments were hallucinogenic or had strong tastes or smells and, at the very least, could have been active or impure placebos.

Brody (1980: 55) described the conditions most conducive to a placebo effect: 1) Patients are given explanations for their illnesses consistent with their preexisting worldviews. 2) A group of individuals assuming socially sanctioned caring roles is available to provide emotional support for a patient. 3) The healing intervention helps the patient to acquire a sense of mastery and control over the illness. The strong resemblance to Torrey's requirements for healing is striking.

Surgery is a particularly strong active placebo because of its powerful symbolism and dramatic performance. Moerman (1983) alluded to coronary heart bypass operations as very successful placebos in our society. The number of bypass operations has dramatically increased since their introduction in the 1960s, reaching an estimated 100,000 a year at a cost of $15,000 to $20,000 each. Numerous studies report successful pain reduction in 80–90 percent of patients with severe angina pectoris, but ventricular function increased in at most only 20 percent of the patients, and no evidence was found for an extension of life except for patients with chronic stable angina and occlusion of the left main coronary artery. The operation also has little rehabilitative effect on patients, who show no improvement on returning to work and do not work longer hours after surgery (Moerman 1983). More interestingly, surgery patients continued to report less pain even after all bypass grafts were themselves occluded, that is, when the operation ultimately failed in its objective. This means, according to Moerman, that the surgery may work but not necessarily for the reasons given, and that it functions as an active placebo (Moerman 1983: 161):

> Bypass surgery, especially, is, from a patient's point of view, a cosmic drama, following a most potent metaphorical path. The patient is rendered unconscious. His heart, source of life, fount of love, wracked with pain, is *stopped!* He is, by many reasonable definitions, dead. The surgeon restructures his heart, and the patient is reborn, reincarnated. His sacrifice (roughly $10,000) may hurt as much as his incisions . . . it seems reasonable to conclude that the general metaphorical effects of this surgery are at least as decisive in its anomalous effectiveness as are graft-patency rates.

Symbolic death and rebirth, without the high technology trappings, is a key component of non-Western healing, including that of the Aztecs.

The scientific evidence for placebo analgesia (a state of not feeling pain though conscious) is fairly clear. Levine, Gordon, and Fields (1978) conducted a double-blind clinical test of pain in dental surgery comparing a local anesthetic and a placebo. The placebo prevented pain and did so by promoting the release of endorphins (endogenous analgesics in the brain, which occupy the same receptor sites as opiates). The placebo analgesic effect could be nullified by administering naloxone, a drug that strips opiate receptors, thus detaching the endorphins. Stress, such as the anticipation of painful electric shock of the foot, elicits the release of endorphins and consequent analgesia in humans (Willer, Deher, and Cambier 1981). Mice exhibit two different responses. Intermittent foot shocks produce opioid analgesia reversible with naloxone, while continuous shocks produce nonopioid analgesia unaffected by naloxone. Only intermittent stress also produces immunosuppressive and tumor-enhancing effects (Terman et al. 1984).

The evidence for immune system effects is less clear, but the new field of psychoneuroimmunology is now a component of behavioral medicine. Stress has been found to have physiological consequences affecting the immune system in mice. Riley (1981) reported a decrease in circulating lymphocytes, shrinking of the thymus, and loss of tissue mass in related areas of the spleen and lymph nodes—all important elements of the immune apparatus. Experimentally, it has been possible to demonstrate in mice stress-associated pathological effects on cancer processes, virus infections, and other diseases subject to immunological control (Riley 1981, Sklar and Anisman 1979). A connection can be traced to the hypothalamus, since stress increased the levels of adrenal corticoids in blood (corticosterone in mice and cortisol in humans), mediated through the hypothalamus and the pituitary.

Data on humans are more ambiguous. The classical study of immune system activity, by Bartrop et al. (1977), found that lymphocytes two months after bereavement of persons who had lost a spouse were significantly less responsive than those of controls. A similar prospective study of men whose wives were terminally ill from breast cancer also found that the husbands' lymphocyte responsiveness dropped after their bereavement (Schleifer et al. 1983). Meyer and Haggerty (1962) reported that chronic stress was related to the incidence of streptococcal infection. Similar studies were summarized in the *Lancet* (Anonymous 1985) and in Morton (1987). Rozanski et al. (1988) showed that mental stress such as that accompanying mental arithmetic or simulated public speaking can trigger episodes of myocardial ischemia (inadequate supply of oxygen to the heart muscle) similar to that induced by physical exercise in people with coronary artery disease, suggesting an explanation for the well-known link between mental stress and heart attacks. These research findings are still controversial; in an editorial Angell (1985) denied any mind-body effect or influence of stress or other psychological factors on the survival rate of cancer patients.

However, such denials by Angell and others may be the last stand of the extreme reductionist wing of Western biomedicine, since evidence continues to accumulate that symbols, and thus the mind, can affect physical health. We should keep the preceding sections in mind as we discuss Aztec magical practices. They may well have been effective as active placebos because they conformed to the worldview of both the sick and the healers and often had strong mythical significance.

As mentioned by Eliade, the curing power of medicines is enhanced by reference to creation myths. Relevant magical examples are found in Aztec medicine. Pustules (*nanahuatl*) were cured by eating the flesh of hummingbirds (Sahagún 1950–1969: book 11, 24) or gold (Sahagún 1950–1969: book 11, 234). In both cases, the mythical reference was to the sun. The disease name evokes the minor god Nanahuatzin ("The Little Pimply One") who sacrificed himself to become the sun in the Fifth Creation. Hummingbirds were related to the sun because Huitzilopochtli ("Hummingbird-on-the-Left"), the patron god of the Mexica, was also a solar god, and because hummingbirds were considered to be the reborn teyolia of warriors who had accompanied the sun in its daily travels. Hummingbird flesh would magically cure nanahuatl because both treatment and diseased were related to Nanahuatzin, that is, the sun. Gold, called *costic teocuitlatl* ("yellow god excrement"), was also considered to be related to the sun. Hummingbird flesh also had two other, unexplained, abilities: to cause sterility (Sahagún 1950–1969: book 11, 24) and to cure epileptics (Hernández 1959: vol. 2, 331).

The cure for bone fractures in Ruiz de Alarcón (1982: 267–270), also found in Hernández (1959: vol. 1, 320), consisted of a plaster made with *poztecpatli* ("fracture medicine"), a splint, and the following incantation:

> Well now,
> O Quail,
> O One from the Place of Disturbance,
> What harm are you doing
> To the bone of the Land of the Dead [Mictlan],
> Which you have broken,
> Which you have smashed?

which continues later with:

> I am the Priest,
> I am the Plumed Serpent [Quetzalcoatl],
> I go to the Land of the Dead [Mictlan],
> I go to the Beyond,
> I go to the Nine Lands of the Dead [Mictlan];
> There I shall snatch up
> The bone of the Land of the Dead [Mictlan].
> They have sinned—
> The priests,
> The dust-birds;
> They have shattered something,
> They have broken something.
> But now we shall glue it,
> We shall heal it. (Ruiz de Alarcón 1982: 267–268)

This incantation refers to the myth, described previously, in which Quetzalcoatl went to Mictlan to retrieve the bones of the first men in order to recreate life. While the god was escaping, the bones were broken by quail sent by Mictlantecuhtli, which explains the reference to quail as culprits in the bone fracture. The healer entered mythic time and became Quetzalcoatl himself, returning to the nine levels of Mictlan to invoke supernatural help for the cure. Andrews and Hassig (Ruiz de Alarcón 1984: 371) saw a further meaning:

> The myth is an explanation of human fragility and perishability. Human mortality is owing to imperfections in the very framework, the hardest part of the body. Mictlan Teuctli's perfidious action is motivated by his realization that if man were made from perfect material, he would never die. By putting his mark

on the bones, he brands them as his property so that they must return to him. Quetzalcoatl's weeping comes at the recognition that, in creating life, he will be creating death. By naming Mictlan as the source of the bones, the myth presents mortality as implicit in human life.

The Aztec treatment of bone injuries was nevertheless effective, possibly the most advanced aspect of Aztec surgery, according to Viesca Treviño (1984b: 214). They used traction and countertraction to reduce fractures and sprains and splints to immobilize fractures (Sahagún 1950–1969: book 10, 153; 1956: vol. 3, 182). They also treated complications such as swelling around the break, incising it with an obsidian lancet or applying a mixture of plants as a plaster. For failure of the bone callus to consolidate in fractures "the bone is exposed; a very resinous stick is cut; it is inserted within the bone, bound within the incision, covered over with the medicine mentioned" (Sahagún 1950–1969: book 10, 1953). Viesca Treviño (1984b: 215) called this "a simple and unpretentious description of an intramedullar nail—a technique not used in Western medicine till the twentieth century."

The above descriptions of fracture treatment by Sahagún and Ruiz de Alarcón are an example of the differences between the two sources. Sahagún presents, particularly in this chapter, Aztec rational medicine without much magic or myth. Ruiz de Alarcón dwelled on the magical and mythical aspects with little in the way of rational, objective practices. The true picture was probably a combination of both. Another example is the treatment of scorpion stings. In Sahagún (1950–1969: book 11, 87) the cure was to suck out the poison and rub tobacco on the wound, while in Ruiz de Alarcón the cure invoked a creation myth. The myth told that in the beginning a man called Yappan, who was destined to become a scorpion, went to live on a rock in abstinence and chastity as penance to mitigate his fate. Some goddesses foresaw that if he persisted in his chastity and were still converted into a scorpion, he would kill all those he stung. They decided to send the goddess Xochiquetzal to seduce Yappan and end his abstinence. She climbed upon the rock and, covering both of them with her skirt, seduced him. Yappan was beheaded and became a scorpion but, because he had sinned, could not kill all those he stung (Ruiz de Alarcón 1982: 293–295). The myth-based cure for scorpion stings was primarily a reenactment of the seduction of Yappan, in which the curer "became" Xochiquetzal. A male curer threw his blanket over the victim; a woman used her skirt and reminded Yappan of her dominance over him (Ruiz de Alarcón 1982: 295–299).

We must remember that European medicine at that time and later also relied extensively on magical treatments. Tuberculosis, for example, has a long history of magical cures. Galen prescribed wet-nurses' milk. Many years later, Sydenham also prescribed wet-nurses' milk and added horseback riding and breathing the air from stables. The use of wet-nurses' milk continued until the end of the eighteenth century. The cure of scrofula by the touch of the kings of France and England was introduced by Clovis in A.D. 496 and was practiced as late as the nineteenth century (Ackernecht 1965: 108). In the seventeenth century, Serna (1953: 97) reported immersing in water a small piece of bone from Saint Gregorio López to cure a very sick woman, whereupon, he said, she vomited a piece of wood containing coal, burnt eggshells, and hair, which had caused the illness.

The existence of such European beliefs must be taken into account, for example, in considering descriptions of the Aztec's magical use of gems to cure disease because their origin might really be European. Hernández (1959: vol. 2, 412), in citing *costic-tecpatl* (carnelian) as supposedly good for the heart, said:

> This property which, as well as those for all other stones, the Mexicans say they have learned from the Spanish, because, prior to their arrival, they were used only as objects of adornment and luxury similar to gold and silver.

In fact, practically all descriptions of the medical uses of gem stones are found in Hernández and in the *Badianus Codex*, which is suspect for European influence. The single exception is *eztetl* ("blood stone," jasper) for which Sahagún gives a personal testimonial. Moriarty (1974) supported the suspicion of European origin by pointing out that the name jade derives from *piedra de ijada* ("stone of the side"), and that the name nephrite, for a variety of jade, was taken from the Greek word *nephros*, for kidney, because jade was believed to cure kidney ailments. He attributed the first description of this use to Monardes, a physician of Seville, in 1565. Hernández (1959: vol. 2, 410, 411) listed two other stones that cured kidney pain—*tlapaltehuilotl* (amethyst) and *itlilayo teoquetzaliztli* (mexican jasper)—which were tied to the side of the body or to the forearm on the same side as the ailing kidney.

Monardes (1925:44–45) said of jasper (Elizabethan English translation by John Frampton):

> The Blood stone is a kinde of jasper of divers coullours some-
> what darke, full of sprincles, like to blood, beeying of coullour
> redde. . . . The use thereof, bothe here and there, is for all fluxe
> of blood, of what partes so ever it bee. . . . The stone must be
> weate in colde water, and the sicke manne muste take hym in
> his right hande, and from tyme to tyme weate him in cold water.

This sympathetic magic use of jasper because of its blood-spattered look is also found in Hernández (1959: vol. 2, 411). The *Badianus Codex* (de la Cruz 1964: 217, 219) cited its use to stop menstruation and postpartum bleeding. Sahagún (1950–1969: book 11, 228) stated that eztetl was used for nosebleeds and described a personal experience:

> I have one more or less the size of a fist. It is rough as if just
> quarried. This year of 1576, this stone has given life to many
> whose blood and life was gushing from their nostrils during
> this epidemic. If taken by the hand and held clenched for a
> while, the blood stopped and they recovered from this illness
> from which many die and have died in this New Spain. There
> are many witnesses to this in this town of *Tlatilulco de Santiago.*
> (Sahagún 1956: vol. 3, 337)

The *Badianus Codex* consistently recommended a group of stones for their cooling properties to treat problems with heat. Except for pearls, which have an obvious aquatic connotation, all stones in the group are green. Heat in the heart (de la Cruz 1964; 179) was treated with among other things beryl, emerald, pearl, and *xiuhtomolli* (turquoise). Fevers were treated with emerald and xiuhtomolli (de la Cruz 1964: 197). The burning pain of gout was treated with pearl and emerald (de la Cruz 1964: 187). The identity of the stone called "emerald" (*smaragno*) in de la Cruz is problematic. Viesca Treviño and de la Peña (1974) believed it was *chalchihuitl* (turquoise), but it may well have been jade, for which there was no Latin term at the time. They also suggest from the etymology of its name that its cooling properties were due to its relationship to Chalchiuhtlicue, the Goddess of Water. In fact, green stones may have been generally believed to attract moisture and therefore be cooling, as the following illustrate:

and thus do they know that this precious stone is there: [the herbs] always grow fresh; they grow green. They say this is the breath of the green stone [chalchihuitl], and its breath is very fresh. (Sahagún 1950–1969: book 11, 222)

Quetzalitztli [emerald-green jade] . . . the genuine [jade] attracts moisture . . . becomes wet, has dew (Sahagún 1950–1969: book 11, 222)

Quetzalchalchihuitl (jade of uniform green color] . . . and its body is as dense as the green stone. Also it is an attractor of moisture. It attracts, it exudes moisture (Sahagún 1950–1969: book 11, 223)

chalchihuitl [common jade of green and white color] . . . Its appearance is herb-green, like the amaranth herb. Also it is one which attracts moisture. (Sahagún 1950–1969: book 11, 223)

Examples of sympathetic magic (disease and cure having similar origins or features) are more common than those of contagious magic (cure transmits desired qualities). Additionally, the magical uses of animals (and gems) are proportionally much more numerous than of plants. Viesca Treviño and de la Peña (1974) agreed and suggested that this may be because behavioral and physical characteristics are more easily observed in animals and may be missed in plants. However, the problem may well be a lack of the botanical identification needed to determine the chemical or morphological characteristics of plants that might rationalize their uses according to the laws of magic. With these general comments in mind, we will give various examples of magical cures. The list is far from exhaustive and can be expanded enormously, particularly by using the Badianus, which is replete with examples. However, its taint of European influence, for example, the purely European use of products like bezoar stones and references to animals like the rooster not native to the New World, makes one reluctant to rely on the Badianus without support from other sources. In any case, Viesca Treviño and de la Peña (1974) gave an extensive and excellent review of magical remedies in the Badianus.

Some cases of sympathetic magic can be analyzed in terms of the morphology of the remedy. For example, *yolloxochitl* ("heart flower," *Talauma mexicana*) is shaped like a heart and was recommended for a variety of heart problems and as a diuretic (de la Cruz 1964: 185, 187, 213; Sahagún 1956:vol. 3, 330). As discussed below, this plant also has active ingredients as well as a suggestive shape. *Hueynacaztli (Enterolobium cyclocarpum)* has large ear-shaped flowers, from which the name derived, and was prescribed for travelers and tired rulers, who presumably had a great need to hear well (de la Cruz 1964: 193, 217; Viesca Treviño and de la Peña 1974). *Tzozoca xihuitl* ("wart herb," *Euphorbia helioscopa*), with leaves full of growths, was a treatment for warts (de la Cruz 1964: 211). Hernández (1959: vol. 2, 406) described algae called *tapachpatli* ("sea shell medicine") that resembled mashed liver and was used to treat liver ailments.

In other cases, the sympathetic magic stems from other properties inherent in the remedy. Plants that exude milky latex, for example, *chichioalpatli,* or "breast medicine" (Hernández 1959: vol. 1, 188) and *cuetlaxochitl,* or *Euphorbia pulcherrima* (Hernández 1959: vol. 1, 319), were applied locally to increase the flow of milk of nursing mothers and wet nurses. *Cocoztamal* ("yellow tamal"), which has a thick yellow root, was used as a diuretic. Hernández related that the Bishop of Cordoba, confessor to King Phillip II, suffered a serious case of urinary retention that was miraculously cured by cocoztamal provided by an Indian doctor (Hernández 1959: vol. 1, 217). *Cuauhalahuac* ("slippery tree," *Grewia terebinthacea*) was used in the birth process in the hope that its slippery qualities would "grease the skids" for delivery of the child (de la Cruz 1964: 217).

Neyoltzayanalizpatli ("medicine for the breaking of the heart") was described by Hernández as glutinous. Its leaves were dissolved in cold water and given to people who had been "frightened," that is, those whose hearts had been damaged by phlegm in the chest. Presumably, the sticky latex would glue the heart together again (Hernández 1959: vol. 1, 70).

Magic related to properties other than shape was more common in animal remedies. Sahagún (1950–1969: book 11, 48) said that the dove, *cocotli,* cried when its mate died, presumably indicating fidelity, and "it is said that it destroys one's torment and affection. They make the jealous eat of its flesh; thus they will forget [their] jealousy." Hernández (1959: vol. 2, 328) stated that a woman can be cured of jealousy by unknowingly eating a cooked dove's flesh, but was at a loss to explain why. A worm called *tlalomitl* ("ground bone"), which was white, firm, and very rigid, was recommended for men who were impotent because it would promote or prolong an erection (Sahagún 1950–1969: book 11, 92). It was also said that the *axolotl (Ambystoma tigrinum)* menstruated like a woman and its flesh was an aphrodisiac (Hernández 1959: vol. 2, 390).

The cure for pain in the joints given in the Badianus presents an interesting combination (de la Cruz 1964: 203):

> Against pain in the joints. A poultice of the following herbs shall be prepared: *cuauhtzitzicaztli, tetzitzicaztli, colotzitzicaztli, patlahuactzitzicaztli* and *xiuhtlemaitl*. Also include small snakes, scorpions, and centipedes, grind everything together and cook in water.
>
> Further, the part affected by stiffness should be lanced with the bone of an eagle or a lion and afterwards the aforementioned plaster mixed with honey should be applied.

The flexibility of small snakes and the many movable joints of centipedes contributed to alleviation of symptoms such as stiffness through sympathetic magic. However, the herbal poultice represented an empirical approach. The herbal names contain the root *tzitzicaztli*, meaning "nettle," and have been identified (de la Cruz 1964: 267–268) as: cuauhtzitzicaztli (*Urera* sp.), tetzitzicaztli (*Cnidoscolus* sp.), colotzitzicaztli (*Cnidoscolus* sp.), patlahuactzitzicaztli (*Tournefortia hirsutissima*), and xiuhtlemaitl (*Pectis* sp.). Three of these plants are extremely irritating and produce welts on the skin due to the presence of histamine and formic acid (von Reis Altschul 1973:162; Gibbs 1974:vol. 1, 363, 422). *Tournefortia hirsutissima* is very aromatic, and *Pectis* smells strongly of lemon (von Reis Altschul 1973: 241, 390). Nettles have been applied locally to treat rheumatism for a long time in Western folk medicine (Thompson 1978:182) and in the Middle East (Boulos 1983: 191). Presumably the induced heat and stinging act as a counterirritant anodyne. No evidence proves it an effective treatment for rheumatism (Tyler 1987:164) but, at minimum, it must have been a very active placebo.

Sahagún, in an early text, also mentioned lancing of the joint: "FORMATION OF VISCID FLUID IN THE KNEES. It is cured by bleeding with a pointed thorn. Something comes out that is akin to the viscid fluid of the nopal [*Opuntia* sp.]" (López Austin 1972: 137). Alarcón-Segoiva (1980) said that this is an accurate description of therapeutic arthrocentesis to relieve abnormal accumulation of fluid in a knee and would be an effective treatment. Cut leaves of the cactus nopal produce a viscid fluid, commonly called the "saliva" of the nopal, that floats to the top when the leaves are boiled. Alarcón-Segovia measured the relative viscosity of the fluid to be 19.5, within the range of synovial fluid, which varies from 9.6 in rheumatoid arthritis to 46.0 in ankylosing spondylitis.

The speed and agility of hares and rabbits derive fundamentally from the structure and musculature of their legs, which would make them good candidates as remedies for ailing extremities (Viesca Treviño and de la Peña 1974). Eating the flesh of cooked hare and rabbit was prescribed for "incipient contraction of the knee" (de la Cruz 1964: 189). Gout (*coacihuiztli*) was treated with a variety of ingredients including the "cold" gems described above. In addition, affected parts were rubbed with the blood of hare, rabbit, a small serpent, or a lizard. The serpent would have been used for its name as well as its flexibility, since coacihuiztli means "stiffness of the serpent." Lizards, as discussed above, were also symbols of agility; people born on a day 1-lizard could not be injured by falls. Lizards were also used magically to cure someone who "fell striking his chest." The cure involved drinking a potion of very bitter herbs, drawing blood from a vessel near the heart to prevent future complications, and drinking a mixture of four (a ritual number) raw, ground-up lizards in urine (Sahagún 1950–1969: book 11, 162).

Examples of contagious magic are much scarcer, perhaps because the sources do not provide the referents that would allow us to identify them. People frightened by lightning were anointed with the sap of a tree that had been hit by lightning and with an effusion of ground herbs from the tree's vicinity (de la Cruz 1964: 213). Fright, as well as insanity and congested heart, was also treated with an unidentified stone called *quiauteocuitlatl* ("rain gold"), believed to be created and then driven into the ground by lightning. This illness belonged to the category of phlegm-in-the-chest diseases, which other sources said were treated empirically with plants to evacuate these humors.

Further examples deal with the power of hair (Viesca Treviño 1984b: 270–272), which, as we have seen, derived from its close association with and its role as a protective cover for tonalli. A remedy for sleepiness was to burn some plucked hairs and to inhale the smoke while allowing some of it to enter the ears (de la Cruz 1964: 161). If a baby refused to suckle, which might be a symptom of sunken fontanelle attributed by the Aztecs to the loss of tonalli (see chap. 8), it was fed ashes from its own burned hair as well as quail's liver that had been exposed for a time to the sun in order to accumulate tonalli.

A number of remedies were magical but difficult to categorize. Swallowing skunk spray (the skunk was associated with Tezcatlipoca) was prescribed as a cure for pustules (*nanahuatl*) (Sahagún 1950–1959: book 11, 13, 17), which were associated with the sun and solar deities. Eating skunk flesh or drinking skunk urine was supposedly a cure for syphilis (Hernández 1959: vol. 2, 304). Drinking skunk's blood to cure venereal disease is still reported today (Aguilera n.d.: 34). Smoke from the *ahuehuetl* (*Taxodium mucronatum*) was supposed to help draw out fetuses and afterbirths (Hernández 1959: vol. 1, 47). The *cozolmecatl* (*Smilax mexicana*) was a veritable panacea since it "returned strength simply by being touched . . . brought the dying back to life . . . restored strength notably to those exhausted by sexual excess if they would just lie upon it" (Hernández 1959: vol. 1, 250–251). Finally, Hernández (1959: vol. 1, 76) reported an interesting wound treatment:

> [*atochietl*]. . . . It was discovered that when deer were shot with arrows and felt wounded mortally, they would run toward it [the plant] and eating it would regain their strength and were able to flee more rapidly. It was thus discovered and it was found that it is extraordinarily beneficial for recent wounds particularly if they are poisoned.

═ *Empirical Cures*

The Aztecs also had a considerable stock of empirical knowledge acquired in part from the botanical gardens first established by Motecuhzoma I in 1467, years before the first documented botanical gardens in Europe (Durán 1967: vol. 2, 247–248). Specimens were brought from all over the Aztec empire. The gardens attained a level that amazed the Spanish conquerors, who quartered there as they approached Tenochtitlan (Cortés 1971: 196). Experience in the gardens was reflected in the Aztecs' extensive and scientifically accurate botanical and zoological taxonomy (Ortiz de Montellano 1984). The gardens were also used for medical research, plants were given free to patients on the condition that they report the results, and doctors were encouraged to experiment with the various plants (del Pozo 1967). The Spanish attested to the high quality of native doctors:

> They have their own native skilled doctors who know how to use many herbs and medicines which suffices for them. Some of them have so much experience that they were able to heal Spaniards, who had long suffered from chronic and serious diseases (Motolinía 1971: 160).

Torquemada (1975–1983: vol. 3, 325) mentioned that Aztec battle surgeons tended their wounded skillfully and healed them faster than did the Spanish surgeons. He also described the infinite number of herbs sold in Aztec markets, the skill needed to distinguish between them, and that doctors cured without using mixtures of herbs (Torquemada 1975–1983: vol. 3, 349). Hernández (1945: 86) picked up this theme and strongly criticized Aztec doctors for not using bleeding, an extensive practice of European medicine at the time, and for using single medicines instead of mixtures of ingredients. This trend toward polypharmacy was current in Europe, reaching an extreme in theriac, an antidote for poison and general cure-all composed of more than sixty ingredients. It may explain why the *Badianus*, which was designed to impress the Europeans, described such complex mixtures of remedies.

One area of clear Aztec superiority over the Spanish was the treatment of wounds. European wound treatment at that time consisted of cauterization with boiling oil and reciting of prayers while waiting for infection to develop the "laudable pus" that was seen as a good sign (Forrest 1982). The Aztecs were engaged in practically continuous warfare, had many opportunities to practice their skills, and developed an empirical approach to wound treatment. The fullest description was provided by Sahagún (1950–1969: book 10, 161):

> [for head wound] . . . First the blood is quickly washed away with hot urine. And when it has been washed, then hot maguey sap is squeezed thereon. When it has been squeezed out on the place where the head is wounded, then once again maguey sap, to which are added [the herb] called *matlalxihuitl* and lampblack with salt stirred in, is placed on it. And when [this] has been placed on, then it is quickly wrapped in order that the air will not enter there, and so it heals. And if one's flesh is inflamed, this medicine which has been mentioned is placed on two or three times. But if one's flesh is not inflamed, this medicine which has been mentioned is placed on only once and for all.

Treatment of wounds with maguey (*Agave* spp.) sap was mentioned elsewhere in the *Florentine Codex* (Sahagún 1950–1969: book 10, 145, 146, 148, 162, book 11, 179) and in Sahagún's earlier versions (López Austin 1972:145). A wide variety of chroniclers such as Motolinía (1971: 364), de las Casas (1967: vol. 1, 311), Hernández (1959: vol. 1, 349) and Pomar (1964: 216) described the use of maguey sap to treat wounds. Wound treatment with other saps was described by Acosta (1962: 189–190), Monardes (1925: 158), and de las Casas (1967: vol. 1, 78). The use of maguey sap to treat wounds continued during the colonial period (Esteyneffer 1978: vol. 2, 630, 635) and persists today (Patrick 1977: 93; Díaz 1976b: 235).

The three steps of treatment were: 1) wash the wound with warm, fresh urine; 2) treat with matlalxihuitl (*Commelina pallida*) to stop the bleeding; and 3) dress the wound with hot, concentrated sap obtained from *Agave* leaves, with or without added salt. Treatment was repeated if inflammation indicated infection was present.

In healthy humans, urine is sterile and a good alternative to water of dubious quality on battlefields (Berkow 1977: 715). *Commelina pallida* contains tannins that precipitate proteins and render them resistant to proteolytic enzymes. This produces an astringent effect when applied to intact skin. It is also used medicinally as an astringent and hemostat for burns and abrasions (Claus, Tyler, and Brady 1970: 131). Although this species has not been analyzed extensively, indications suggest that other active compounds are present. Chinese folk medicine, independently of the Aztecs, also used *Commelina pallida* as a hemostat (American Herbal Pharmacology Delegation 1975: 117). As we have pointed out, the independent use of a particular plant species for the same illness by two different cultures enhances the credibility of historical reports.

Early modern reports of the physiological activity of *C. pallida*, although not conclusive because of outdated methodology, are very suggestive. In one study, for example, an extract of *C. pallida* caused intact jugular veins of pigeons to contract until they shut (Instituto Médico Nacional 1894–1907: vol. 1, 368–372). The plant also caused strong hemostatic effects and smooth-muscle contractions that induced pregnant dogs and rabbits to abort. These vasoconstrictor effects as well as the ability of plant extracts to accelerate blood coagulation were also demonstrated by Pérez-Cirera (1944). The narrowing of blood vessels and contraction of smooth muscle cannot be due to tannins but would be consistent with the presence of lysergic acid derivatives such as those found in ergot (*Claviceps purpurae*) or ololiuhqui (*Turbina corymbosa*). Further chemical research would be needed to verify this hypothesis.

Agave sap has a number of components, but saponins and

polysaccharides (approximately 10 percent by weight in dilute sap) are most relevant to its action. Saponins can have hemolytic and detergent effects, which may account for the sap's antibiotic, antiviral, and fungistatic activity (Davidson and Ortiz de Montellano 1983; Tschesche and Wulff 1972). Even in small concentrations these compounds would contribute to the maintenance of an antibiotic presence at the wound site.

The antibiotic effect of maguey sap is due to osmotic pressure, which causes solvent flow between solutions of different concentrations separated by a semipermeable membrane in the direction of equalization of concentrations. The membrane allows water to pass through from the more dilute solution but not solute molecules such as salts or their ions. In a simplified view, bacteria are dilute water solutions of enzymes surrounded by a semipermeable cell wall. If bacteria are placed in a concentrated ionic or sugar solution, they lose water by osmosis and perish from dehydration. Majno (1975: 117–118), investigating the antibacterial properties of honey and grease, an Egyptian wound remedy, found that heat-sterilized honey efficiently killed both *Staphyllococcus aureus* and *Escherichia coli* by osmosis. Hypertonic solutions like honey also assist healing by promoting a flow of serum to the wound site, thus supporting the body's natural defense mechanisms. This effect was found to be particularly beneficial for dirty or infected wounds (Majno 1975: 118; Cope 1958).

Sugar also can produce a hypertonic solution in wounds. Argentinian surgeons Herszage, Montenegro, and Joseph (1980) treated 120 patients with infected wounds by debridement, packing the opening with ordinary granulated sugar, and replacing the sugar as needed. The cure rate exceeded ninety-nine percent, even in the case of difficult wounds or diabetic patients. The therapeutic action was experimentally verified in vitro: typical microorganisms that infect wounds were prevented from growing by immersion in sugar solutions (Chirife et al. 1983). Each microorganism had a characteristic osmotic pressure above which it would not grow. Of the bacteria tested, *Staphyllococcus aureus* was the most resistant to this treatment. Sugar packing has been shown to heal infections following cardiac surgery faster and with fewer complications than conventional treatments (Trouillet et al. 1985).

Davidson and Ortiz de Montellano (1983) tested the ability of commercially available syrup made from maguey sap to inhibit the growth of six common bacteria both gram-positive and gram-negative, found in wounds (*Salmonella paratyphi, Pseudomonas aeruginosa, Escherichia coli, Shigella sonnei, Sarcina lutea,* and *Staphyllococcus aureus*). Undiluted syrup inhibited the growth of all but *S. aureus,* and a combination of maguey syrup and salt worked against all six microorganisms. Thus, several studies have verified that the Aztec treatment of wounds was effective by etic biomedical standards.

Another area of effective Aztec medical methods was obstetrics. At approximately the seventh or eight month, a midwife visited an expectant mother, delivered a homily and instructions (often magical injunctions, discussed in chap. 6), and palpated the child while the mother-to-be lay in a steam bath (Sahagún 1950–1969: book 11, 149–156). Just before delivery, the midwife washed and massaged the mother in the steam bath and performed an external version (turning the fetus by external manipulation) if the fetus was in a breech position (Vargas and Matos Moctezuma 1973). The mother assumed a squatting position for delivery, rather than the lithotomy (lying down) position customary today. Lithotomy and its required pushing can deprive the baby of oxygen (Boston Women's Health Book Collective 1984: 372, 373, 380; Page, Villee, and Villee 1976: 184). A squatting position also reduces tearing of the perineum and the need for episiotomies (Stewart, Hilton, and Calder 1983).

If necessary to hasten birth, the woman was given an oxytocic; *cihuapatli* ("women's medicine," *Montanoa tomentosa*) was universally cited in the early sources (de la Cruz 1964: 217; Hernández 1959: vol. 1, 294; Pomar 1964: 215; Sahagún 1950–1969: book 6, 159, 167; book 11, 180). This herb continued to be used after the Spanish conquest and is still used widely, particularly in rural areas of Mexico (Lewis 1970: 358; Quiroz-Rodiles 1945; Ortiz de Montellano and Browner 1985). Despite the unanimity of early opinion and the persistence of use, scientific evidence of the effectiveness of *M. tomentosa* was mixed for some time. Extracts were found to induce contractions in isolated animal uterine tissues (Derbez, Pardo, and del Pozo 1945) and in women in their second trimester who had dead fetuses (Sentíes and Amayo 1964) but not in rat's uterus (Caballero and Walls 1970). However, Caballero and Walls had not used aqueous extracts, which proved to be the source of the active compounds, and had studied rats, which are not a good biological model for cihuapatli.

Levine et al. (1979) discovered two novel compounds, zoapatanol and montanol, in an aqueous extract from *M. tomentosa* and stimulated a flurry of research that demonstrated the uterotonic and antiimplantation activity of zoapatanol in guinea pigs, mice, rats, and hamsters (Hahn et al. 1981). Attention has since focused primarily on its potential as a contraceptive rather than an oxytocic. Studies of its effect on women in the early stages of pregnancy showed that the drug caused strong uterine contractions and cervical dilatation following a single oral dose (Landgren et al. 1979) and caused no ill effects in volunteers given cumulative doses of up to 124 grams of leaves. Extracts of *M. tomentosa* had no acute or subacute toxicity in animals (Southam et al. 1983). More recent studies showed that neither zoapatanol nor montanol when isolated was uterotonic, but that the principal agent was kaurenedienoic acid (Lozoya and Lozoya 1982: 214–219). However, the aqueous extract is believed to be a synergistic mixture superior to the isolated compounds (Lozoya and Lozoya 1982: 218).

The status of a second oxytocic is like that of cihuapatli prior to 1979: evidence of activity but no scientific proof. Sahagún's informants stated that if cihuapatli failed to induce labor, an infusion of the tail of the *tlaquatl*—opossum (*Didelphis marsupialis*)—was infallible (Sahagún 1950–1969: book 6, 159). This also appeared elsewhere in Sahagún (1950–1969: book 6, 167; book 10, 156; book 1, 181) and in Ruiz de Alarcón (1982: 221). Like cihuapatli, opossum tail was used in the nineteenth century (León 1910: 64) and even today (Martín del Campo 1956; Quiroz-Rodiles 1945; Lewis 1970: 357; Fragoso 1978: 59). Quiroz-Rodiles (1945) demonstrated that an extract from either fresh or dried tails produced contractions in both pregnant women and rodent uterine tissue in vitro. This lead has not been followed up, probably because it seems like the magical use of animals discussed earlier. A possible explanation for the choice of opossum rather than any other animal is that the opossum was the only marsupial mammal known to the Aztecs. Since its offspring were "outside" the uterus, by sympathetic magic it might also help the child to get "outside" the mother.

In various cultures remedies have been chosen for erroneous reasons but found empirically effective and so retained in the pharmacopoeia, for example, mercury to treat syphilis in Europe. In some African tribes fresh seminal fluid is taken orally when labor is delayed, presumably because of hoped-for contagious magic: "what got the baby in can also get it out" (Harley 1941: 237–238). Similarly, severe menstrual cramps in women used to be dismissed as purely psychosomatic until the discovery of prostaglandins which produce both menstrual and premenstrual pains relieved by prostaglandin inhibitors (Gilman et al. 1980: 674, 711). The discovery of prostaglandins makes the use of semen to induce labor logical since seminal prostaglandins taken orally are very effective in stimulating contractions of the pregnant uterus, in both early and late pregnancy. They can function as both abortifacients and oxytocics (Karim 1972:102). These compounds are effective taken orally. The amount of prostaglandins present in a single ejaculate corresponds to the dose used to induce labor at term (Karim 1972:103, 141). Therefore, although not proven, the oral use of semen as an oxytocic becomes scientifically reasonable.

Prostaglandins are also a plausible explanation for the use of opossum tail as an oxytocic. Dawson (1977) pointed out that the tail of the kangaroo (another marsupial) has a complex network of superficial blood vessels involved in heat dissipation. McManus (1969), in studying the response of the opossum to heat stress, found that above 37°C peripheral blood vessels widened, particularly in the tail, feet, nose, and marsupium, and that the opossum would lick its tail to increase heat loss by evaporation. Vessel narrowing appeared to play a major role in heat conservation; the tail then became pale from reduced blood flow, and its surface temperature dropped sharply. Prostaglandins are the agents that mediate the tail's temperature-regulating role in certain mammals, including the opossum. The vasodilating and vasoconstricting ability varies between different groups of prostaglandins and between species. The PGE group in all animal studies acted as vasodilators, while the PGF group acted as vasopressors in man but vasodilators in the cat and rabbit (Karim 1972: 166, 174; Gilman et al. 1980: 673). The chain of reasoning validating tlaquatl as an oxytocic is thus: 1) The oppoosum's tail is involved in temperature regulation, which implies the presence of prostaglandins. 2) Prostaglandins are oxytocic in extremely small doses. 3) Prostaglandins are active in oral administration. 4) Both clinical and laboratory studies indicate uterotonic activity in opossum tail extracts. 5) Therefore the use of opossum tail extracts as an oxytocic is scientifically reasonable and probably effective. The missing link is the direct isolation of prostaglandins from opossum tail in sufficient concentration to cause oxytocic activity, which requires further research.

Other empirical cures were described. Snake and spider bites were treated by making cuts at the site and sucking out the poison, then rubbing with ground-up tobacco (presumably to act as an analgesic), followed by a careful diet (Sahagún 1950–1969: book 11, 77, 87, 88, 91). Plastic surgery was performed, for example, the suturing of lacerated lips with hair; salted maguey sap was applied to prevent infections and help healing. If the wound was slow to heal, it was cut again to form clean edges, and suturing was repeated (Sahagún 1950–1969: book 10, 146). A cutoff nose was sewn back in place with hair, and salted bee honey was used to prevent infection. If the nose fragment was not available, a prosthesis was made as a substitute. Unfortunately, the prosthetic material was not described (Sahagún 1950–1969: book 10, 146). Surgical procedures generally made use of obsidian scalpels. Modern research has demonstrated that the edges of obsidian blades are much thinner than those of surgical steel: the sharper the blade, the less damage to tissue. Obsidian blades would be superior in operations such as cataract surgery in which the thinnest incision is desired. Despite their sharp edges, obsidian blades are quite durable if used only on soft tissue (Anonymous 1981; Robicsek and Hales 1984: 67; Young 1981).

An important component of Aztec medicine was a concern for correct behavior as a way of maintaining health or administering therapy. Some prescriptions related to the physiological consequences of harming or losing tonalli, but many were empirical. For example, people with headaches were told not to sit in the sun, work, or enter hot steam baths (de la Cruz 1964: 153). Hot and irritated eyes were to be shielded from the heat of the sun, the glare of white things, and the irritation of smoke and wind. As usual, the Badianus mixes such advice with various magical prescriptions, such as wearing a red crystal or the eye of a fox (de la Cruz 1964: 157–159). Preventive measures were particularly noticeable in oral hygiene, which Sahagún (1950–1969: book 10, 147) described extensively. Teeth were to be polished with charcoal (a good abrasive) and salt; occasionally tartar was removed by scraping with metal tools followed by further polishing. Unsweetened chewing gum was also recommended to maintain clean teeth. De la Cruz (1964: 165, 171) recommended white ash for cleaning teeth and a mixture of herbs as a mouthwash for halitosis. No magical ingredients were included in these prescriptions. Hernández lists a number of substances to be used as dentifrices (1959: vol. 1, 214, 252, 315, 331; vol. 2, 233, 405, 409–410) and against halitosis (1959: vol. 1, 153, 316; vol. 2, 12, 95). A full study of this aspect of Aztec medicine was made by Fastlicht (1971).

=== *Empirical Validation of Aztec Medicine*

To validate empirically the emic aspects of Aztec medicine, testable hypotheses must be derived from Aztec etiologic ideas. In some cases, these ideas were clearly stated, in others, detective work is necessary to deduce the etiologies, as was done in the attribution of headaches to excess blood in the head. The required cure then can be inferred from the presumed etiology. To evaluate a remedy in emic terms, we must determine whether the proposed remedy could have produced the physiological effect desired by the Aztecs. A correct botanical identification of herbal remedies is essential. To minimize errors in our study, we considered only plants identified as the same by modern studies and appearing in more than one of the original sources. The physiological and pharmacological activities of chemical components of the plant were then compared with the effects required by Aztec etiology.

If the chemical substances can produce most or all of the required effects, the plant was deemed effective in emic terms. If only some of the effects but not others can be produced, or if available information concerns crude extracts[2] but not pure compounds, the plant was considered possibly effective. If the reported constituents cannot produce the desired effects, the plant was considered ineffective. Negative results were explained in four ways: 1) the plant is ineffective; 2) active substances are present but have not been fully studied; 3) the botanical identification is not correct; or 4) the Aztec method of preparation affects the active substances in unknown ways. Emic evaluation of Aztec medicine by these criteria showed that the Aztecs were accurate observers of the physiological effects of plants and could produce whatever effects Aztec theories of physiology and etiology required. Fevers were believed to be due to hot phlegm in the chest and had to be removed to cure the condition. This could be accomplished by a diuretic, a diaphoretic, an emetic, or a purgative. A study of 39 of the 62 fever remedies in Sahagún identified botanically showed that 69 per cent of the species could produce the effects required by the Aztecs to cure a fever (Ortiz de Montellano 1979b). This does not mean that an Aztec diaphoretic cures fevers by modern Western standards, but that if Aztec etiology required sweating, a physician could induce perspiration in a patient. A full listing of these remedies is beyond the present scope, but an abundant sampling is listed in appendix B. The table in appendix B is a revision and expansion of previous work and includes an estimate of the effectiveness of each plant by both emic and etic standards[3] followed by individual discussions.

Because emic evaluation of Aztec medicine showed that the Aztecs were accurate observers of the natural world, they should be expected to have discovered remedies valid by etic Western standards. The etic worth of herbal medicines can be rated as effective, possibly effective, or ineffective on the basis of their known constituents, their physiological effects, and Western biomedical principles. Data are available on enough cases to give some indication of the level of confidence of such predictions. This is qualitatively analogous to the levels of confidence used in statistical inference. We have proposed that four such confidence levels be used in evaluating folk medicinal plants (Ortiz de Montellano 1981; Ortiz de Montellano and Browner 1985; Browner, Ortiz de Montellano, and Rubel 1988).

Level 1. Reported folk use, the lowest level of validation. Parallel reports of use for the same purpose by widely dispersed and different populations, and among whom diffusion is unlikely, increase the probability that the activity is actually present. Similarly, reports of use over long periods of time strengthen the evidence, because one would expect ineffective herbs to have been eventually discarded.

Level 2. Reported folk use *and* isolated compounds or extracts with the desired physiological activity. Drugs at this level show the desired activity either in vitro or in vivo with test animals.

Level 3. Reported folk use, activity in test animals or in vitro, and a plausible biochemical mechanism for the manner in which the plant constituents exert the indicated physiological action.

Level 4. Reported folk use, activity in test animals, a biochemical mechanism for physiological activity, and clinically tested on humans or commonly used in medicine.

The optimum Level 4, controlled clinical trials, has been achieved for only a few of the medicines used by the Aztecs. However, many of the drugs and procedures common in Western medicine also have not been subjected to this level of validation (Office of Technology Assessment 1978:93–94; Ortiz de Montellano

1980b). In some cases even a level 3 or level 2 proof is not possible for a remedy that may actually be effective because the active ingredient may not yet be known to science. For example, the Thongan custom of ingesting human semen to induce labor (Junod 1927: 40) and the popular American belief that sexual intercourse late in pregnancy can cause premature labor were dismissed as superstitions until prostaglandins were discovered in semen in sufficient concentration to bring about contractions of the uterus (Karim 1972: 141).

The estimates of effectiveness of the Aztec herbal remedies listed in appendix B were based on the criteria for the four levels described above. Of the 118 different medical applications tabulated for the various plants, 104 (84.7 percent) are effective, in terms of Aztec etiology, 10 (8.5 percent) are not effective, four (3.3 percent) are either due to magic or uncertain. As expected, the results are less successful in etic terms. However, 60 percent are still effective at level 2, 3, or 4 and approximately 18 percent at level 1, and only 22 percent are not effective. Thus, the Aztecs clearly were keen observers of nature and used empirically derived remedies.

As acknowledged above, Aztec medicine was really a mixture of natural and supernatural elements, separated here to simplify analysis. A truer and more complete picture shows the mixture, as in this prescription to help women give birth:

> drink a medicine made from the bark of the tree *cuauhalahuac* and of the herb *cihuapahtli,* ground up in water with a stone called *eztetl* and the tail of the little animal called *tlacuatzin.* She should carry the herb *tlanextia* in her hand. Hair and bones of a monkey, the wing of an eagle, a bit of the tree *quetzalhuexotl,* deer hide, cock and rabbit gall, and onions dried in the sun should be burnt. Salt, a fruit called *nochtli* and *octli* should be added to all this, (de la Cruz 1964: 217)

In this case, all elements are called into play. Cihuapatli (*Montanoa tomentosa*) and tlacuatzin (opposum tail) are scientifically proven oxytocics. Cuauhalahuac ("slippery tree") lubricated delivery by sympathetic magic, and eztetl (jasper) was supposed to stop bleeding magically. The use of deer and monkey parts refers to the days 1-deer and 1-monkey in the book of days (tonalamatl) during which the cihuateteo, women who had died in childbirth, might return to earth. Eagles were symbolically related to war: The most skilled warriors belonged to the Order of the Eagles, and the hearts of sacrificed victims were called *quauhnochtli* ("eagle prickly pear fruit"). Birth was considered symbolic combat, so the religious metaphor of the eagle was appropriate. One-eagle also was a day in which the cihuateteo descended. Incantations similar to those recorded by Ruiz de Alarcón were undoubtedly recited, thus covering all therapeutic bases.

Syncretism in Mexican Folk Medicine

Before the Conquest, pregnant Aztec women avoided going out at night lest they see a lunar eclipse. An eclipse was thought to be due to bites taken out of the moon, and a baby born to a woman who had witnessed a bitten moon might by sympathetic magic have a harelip, that is, a bite taken out of its mouth. For protection, a pregnant woman would place an obsidian knife next to her belly before going out at night (Sahagún 1950–1969: book 7, 8). I have told this story in places far removed from Mexico and the Aztecs (Salt Lake City 1972, Denver 1976, San Antonio 1979 and Detroit 1984) and found the same belief today, in all its details, in the Chicano community. The sole variation was the substitution of some other metallic object, usually a key, for the obsidian knife. The extraordinary fidelity of transmission of this belief is due in part to its relation to a crucial event in a woman's life and its susceptibility to direct oral communication from mother to daughter, wholly outside any system of formal education.

Two cultures, two religions, and two kinds of medicine met in Mesoamerica. Historians and anthropologists have debated about the ways in which Aztec medicine and sixteenth century European medicine then mixed and left traces in Mexican folk culture. Some, notably George Foster, see only European influence in such areas as the "hot-cold" concept of illness and in some culture-bound syndromes. Others, including the present author, think that Mesoamerican beliefs and practices retain many elements of Aztec medicine and the Aztec worldview and, indeed, are an important source of information about Aztec medicine. Although untangling different cultural contributions to modern folk medicine is difficult, the task is worth attempting in detail, to gain a better understanding of Aztec medicine in particular and more generally to delineate the process of syncretism in Mesoamerica. The characteristics of the Aztec worldview and religion that facilitated syncretism, particularly in religion, have already been discussed. Modern Mexican folk medicine is also a syncretic mixture of European and pre-Columbian beliefs.

═ Tlaloc's Herbs

One way to illuminate the process of syncretism is to follow one particular aspect through the years from pre-Columbian times to the present. This brings a historical perspective to current practices that might otherwise seem entirely European in origin. We shall focus on two plants: *yauhtli* (*Tagetes lucida*), the "african day flower" also known as *pericón* and *flor de Santa María;* and *iztauhyatl,* or wormwood (*Artemisia mexicana*), also known as *estafiate* and *ajenjo del pais.* These plants have been associated with and used as remedies for diseases "caused" by Tlaloc and other deities in the rain-agriculture-fertility complex for at least 500 years (Ortiz de Montellano 1980a) and with no other gods. We shall, first, demonstrate that these two plants were indeed related to the deities of the Tlaloc complex, second, show that they have been used to treat diseases associated with these deities, and third, discuss the process of syncretism during the Colonial period in the medicinal applications of these plants. Finally, modern uses of these plants will be analyzed in terms of the historical pattern.

The deities in Tlaloc's complex are particularly significant. The Spanish Conquest shattered the ideological hegemony of the Aztec Empire, and many Indian communities, particularly those isolated by geography or language, had little interaction with others. In this situation, concepts important to farming, particularly concerning rain, were retained much more than those related to the Aztec warrior elites or to the esoteric details of religious practices (Madsen 1967: 372). These retained concepts were in turn often syncretized with European and Catholic beliefs. Tlaloc gained importance, subsumed many characteristics of other deities, and became the prime ingredient of the resulting mixture.

Tlaloc's relative primacy was evident in rituals that took place during months in the Aztec calendar. The following are examples of rituals associated with Tlaloc.

The Atlcahualo ("cessation of rain") ceremony, first month, sought to ensure rainfall during the planting of crops. Rituals included the sacrifice of little children (preferably those called "human banners" because they had a double widow's peak, a mark of selection by the god) on the crests of certain mountains. The *tlaloque* and the *tepictoton,* dwarflike assistants of Tlaloc, lived in caves in the mountains and brought forth rain. Mountains were believed hollow and filled with water for this purpose. One of the tlaloque listed by Sahagún's informants in the *Primeros Memoriales* (Sahagún 1906: fol. 263v) was Yauhqueme ("dressed in *yauhtli*"). His dark green hat was said to be yauhtli-colored. During Atlcahualo, a child, called Yauhqueme for the occasion, was sacrificed atop a hill of the same name.

The Tozoztontli ("small vigil"), third month, was dedicated to Tlaloc and corn deities, but the sources do not explicitly mention *iztauhyatl* or *yauhtli*.

The Etzalcualiztli ("eating corn and bean stew"), sixth month, celebrated the rains and prayed for a good harvest. In one ceremony sacrificial victims were drowned in honor of the deity. The principal priest carried a bark-paper incense bag adorned with sea shells and full of powdered dried yauhtli (Sahagún 1950–1969: book 2, 80). At the end of the month came ceremonies for Chalchiuhtlicue, the water goddess and female counterpart of Tlaloc. The priests scattered yauhtli at her feet (Sahagún 1950–1969: book 2, 21). On another occasion, the priests painted themselves blue, in a pattern characteristic of Huixtocihuatl, the salt goddess and a member of the Tlaloc complex. They carried bark-paper incense bags in the shape of aquatic birds and filled with powdered yauhtli. The powder was scattered as an offering and burned as incense. Victims were sacrificed to Tlaloc, and spectators carried flowers of iztauhyatl. Little children fanned with the flowers were believed to resist worm infestations (Sahagún 1950–1969: book 2, 82–83).

The Tecuilhuitontli ("little feast of the lords"), seventh month, honored Huixtocihuatl. A description of part of her costume stated: "And her reed staff was hung with papers; it had papers. And they were spattered with liquid rubber; they had rubber. In three places [the staff] had cuplike flowers . . . in each one there was yauhtli" (Sahagún 1950–1969: book 2, 92). Female salt makers, who were her proteges, honored her with dances for ten days, carrying a rope with interwoven iztauhyatl flowers and wearing garlands of the flower. Victims, called *uixoti,* who impersonated the goddess, were painted and dressed like her. They were sacrificed at the temple of Tlaloc where the spectators also carried iztauhyatl flowers and *cempoalxochitl (Tagetes erecta),* a plant of the same genus as yauhtli.

The Tepeilhuitl ("feast of the hills"), thirteenth month, was dedicated to Tlaloc and to the tlaloque, but no specific mention was made to yauhtli and iztauhyatl.

The Atemoztli ceremonies ("the rain falls"), sixteenth month, honored Tlaloc and sought to ensure that the rains would come with prayers to Tlaloc accompanied by repeated offerings of burning yauhtli in the four cardinal directions. Amaranth dough images of the hills inhabited by tlaloque, Yauhqueme among them, were made, and food offerings were placed before them (Sahagún 1950–1969: book 2, 151–152). As we have seen, these dough images also had curative properties.

Yauhtli was mentioned several other times in connection with members of the Rain God complex. Opochtli, one of the tlaloque, was the patron of boatmen and fishermen. Ceremonies in his honor involved offerings of yauhtli (Sahagún 1950–1969: book 1, 37). A list of the duties of priests included the provision of yauhtli for ceremonies honoring the goddesses Xilonen and Tzapotlatenan (Sahagún 1950–1969: book 2, 208–209). Tzapotlatenan was a member of the Rain God complex, while Xilonen, a corn agricultural goddess, was closely related to rain deities.

Apart from one case,[1] all uses listed for yauhtli and iztauhyatl in the *Florentine Codex* involved ceremonies associated with Tlaloc and gods in his complex. In prayers, yauhtli was sometimes addressed as if it were the god himself (Ortiz de Montellano 1980a; 1986). Given the interconnections that Aztecs saw between religion, the natural world, and human fates, it is not at all surprising that plants associated with a particular deity should play a role in curing ailments also associated with the deity and the deity's attributes. In many cases, the chemical or pharmacological properties of the plants suggest the association; in others, the relationship seems to be magical. Some god-related diseases have been mentioned previously; table 8.1 presents a summary, giving the Nahuatl name, a literal translation, the equivalent disease in European terms, and whether the death was sent by Tlaloc and/or sent or caused by possession by such assistants as the tlaloque, tepictoton, ahuaque, or chaneque (spirits that lived in springs and cold caves).

Many of these ailments, and others, were treated with iztauhyatl and yauhtli. Such diseases can be grouped according to their supposed etiology and proposed cure in order to better understand the ideology underlying the treatment. The illnesses listed in table 8.2a, for example, were all believed due to either the slow accumulation of hot phlegm in the chest or its sudden accumulation because of a terrifying experience such as a narrow miss by lightning, as previously described. López Austin (1988: vol. 1, 337) agreed with the grouping of theses diseases but felt that the Aztecs imputed them to possession by ahuaque, the teyolia of people who had died by Tlaloc's will.

Yauhtli and iztauhyatl were cures in two senses. They were Tlaloc's herbs used for Tlaloc's diseases but also had rational physiological effects. The logical remedy for an excess of phlegm is its expulsion via perspiration, feces, urine, or vomit. Yauhtli contains the flavones quercitagritin and kaempferitrin (Hegnauer 1962–1977: vol. 3, 526, 528). Quercetin glycosides are diuretic, while kaempferitrin and its glycosides are laxatives (Steinegger and Haensel 1963: 169, 174).

The diseases listed in table 8.2b seem to have an element of divine causation combined with exposure to cold. The connection is clear in the case of *teococoliztli* ("divine illness"), usually translated as leprosy by the Spanish. As shown in table 8.1, coacihuiztli was associated with punishment by Tlaloc but was also believed to be caused by winds that were cold or issued from caves, thus involving both divine and natural (physical) origins. Iztauhyatl and yauhtli as cures were both divine intercessors, due to their close connection to Tlaloc, and appropriately "hot" counteragents for illnesses caused by cold. Since "hot" and "cold" are somewhat metaphorical qualities, they cannot always be correlated with chemical compositions. Later in this chapter we will examine more closely whether the Aztecs actually did classify plants as "hot" and "cold."

Table 8.2c lists diseases apparently with primarily physical

Table 8.1. *Ailments Associated with Tlaloc and the Rain God's Complex*

AZTEC DISEASE	TRANSLATION	EQUIVALENT	TLALOC	HELPERS	REMARKS
Teococoliztli	god's disease	leprosy	X	—	breaking fast of Atamalcualiztli
Coacihuiztli	stiffening of the serpent	gout, paralysis, stiffness	X	X	associated with cold, wet, violation of pulque-making ritual
Atemaliztli	swelling due to water	edema, buboes, swelling	X	—	phlegm in chest
Atonahuiztli	aquatic fever	intermittent fever (malaria?)	—	X	phlegm in chest, possession via lightning
Aacqui	lightning struck, intrusion	epilepsy, madness, insanity	—	X	narrow miss by lightning or strong wind (*susto*?)
Netonalcahualiztli	abandonment of tonalli	loss of tonalli due to fear	—	—	*susto?*
Mocuitlapam-mauhtique	she whose back is terrified	*susto* due to lightning	—	—	
Netlahuitequiliztli	struck by lightning	death by lightning	X	—	also sent by Chal-chiuhtlicue

Source: References in Ortiz de Montellano 1986. Reprinted with permission from R. P. Steiner, ed., *Folk Medicine: Art and the Science.* Copyright © 1986 by the American Chemical Society.

Table 8.2. *Pre-Columbian Uses of Iztauhyatl and Yauhtli*

DISEASE	Iztauhyatl	Yauhtli	Etiology	Cure
			(a) DISEASES CAUSED BY PHLEGM	
Atonahuiztli (intermittent fever)	X	X	phlegm in chest	rational: expel phlegm
Aftereffects of wind and lightning	X	X	"	"
Fevers, excess phlegm	X	X	"	"
Madness, epilepsy, ailments from oppression of the heart	X	X	"	"
			(b) DISEASES DUE TO DIVINE CAUSES	
Coacihuiztli (gout) Paralysis, stiffness	X	X	divine plus "cold"	religious plus "hot"
Teococoliztli (leprosy)	X	X	divine	religious
Relapse from illness	X	X	divine plus "cold"	religious plus "hot"
Spitting blood		X	?	?

(continued)

Table 8.2. (Continued)

DISEASE	Iztauhyatl	Yauhtli	Etiology	Cure
		(c) DISEASES DUE TO NATURAL CAUSES		
Dirty kidneys, diuresis	X	X	excess water	diuretic
Swellings, blisters	X	X	"	"
Thirst of dropsy		X	"	"
Pains of cold origin	X		"cold"	"hot"
Digestive troubles (empacho), children's colic	X	X	"	"

Source: Ortiz de Montellano 1986. Reprinted with permission from R. P. Steiner, ed., *Folk Medicine: Art and the Science.* Copyright 1986 by the American Chemical Society.

Table 8.3. *Magical Uses of Iztauhyatl and Yauhtli*

USE	SOURCE	PLANT	MODERN USE
Tetlacuicuilique *(object extraction)*	a	iztauhyatl	yes (limpia plus "hot" herbs)
Techichiniani	a	iztauhyatl	yes (limpia plus "hot" herbs)
Prevent worms in children's eyes during Etzalcualiztli	b	iztauhyatl	no
Crossing a river safely	c	yauhtli	no

Source: a. Garibay (1946); b. Sahagún (1950–1969: book 2, 82); c. de la Cruz (1964: 215).

causes, such as an excess of either water or cold, both characteristics expected of a rain god. Rational cures would be a diuretic and a "hot" medicine. Yauhtli and iztauhyatl actually contain both diuretics that promote water excretion and diaphoretics that manifestly induce "heat" and perspiration (Ortiz de Montellano 1974/75; 1975).

Iztauhyatl and yauhtli were also used to cure the magically caused ailments listed in table 8.3. Many cultures, including the Aztec, have believed that certain diseases result from the introduction of small objects such as pieces of bone, flint, or pebbles under the victim's skin. The remedy usually involves removal of the offending object by suction; the Aztec healer first chewed iztauhyatl and sprayed it on the patient's skin. The connection with the water connotation of Tlaloc and his complex was not explicit, but the rationale may simply have involved expected benefits from the god's close association with the herb.

The other two tabulated uses of these herbs were clearly connected to Tlaloc. In one case, the herb was a magical charm that enabled a person to cross a river safely; the other occasion was a festival dedicated to Tlaloc in which the herb was a magical shield against parasitic infections. Although object intrusion is no longer regarded as a major source of illness in Mexican or Chicano folk medicine, the magical power of iztauhyatl is still exploited. The herb is an important component in frequent ritual cleansings (*limpias*), particularly for the ailment called *susto* (Garibay 1943–1946). Syncretism is evident in the frequent inclusion in limpias of two herbs of European origin—rue (*Ruta graveolens*) and rosemary (*Rosmarinus officinale*). These herbs were classified as "hot" in the Hippocratic system, just as iztauhyatl and yauhtli were "hot" in the Aztec system. As discussed below, the European herbs have apparently become the generic equivalent of the Aztec plants.

═ Colonial Period

Much of the following analysis is based on the seminal work of Gonzalo Aguirre Beltrán (1963) on the process of acculturation in Colonial Mexico. The cultural shock and rapid changes caused by the Conquest and the imposition of a new religion created enormous tensions. Mixed racial groups such as the mestizo and the mulatto were relegated to a low social status, and their culture resembled that of the Indian mother (since relatively few Spanish women migrated) rather than that of the Spanish father. They were marginal people, straddling two cultures and trying to reconcile and reduce the friction between two worldviews. Such efforts of accommodation, a similarity between Catholic and Aztec rites, and a compatibility of certain medical concepts (such as maintenance of equilibrium and moderation to promote health) opened the way to syncretism.

As the Catholic church and the Inquisition intensified efforts at conversion and the extirpation of all traces of previous indigenous beliefs, many native cultural and religious tenets went underground. They survived surreptitiously, although sometimes altered as time passed and old men educated in the *calmecac* before the Conquest died out. Just as belief in Tlaloc survived the rise and fall of other native empires, before the Spanish came, because the god was closely associated with rainfall and agricultural fertility, his worship persisted during fragmentation of the Aztec Empire by the Spanish. The more extensive and complex pantheon of warrior and calendric gods did not survive the loss of trained priests (Madsen 1967: 372). Aguirre Beltrán attributed cultural preservation principally to the nahualli (shaman, described in chap. 2):

> in the past, the *nagual* was entrusted with the task of resolving the anxiety due to the hazards of rainfall agriculture in a semi-desert ecology. From birth the life of a *nagual* was ruled by the day sign 1-*quiahuitl*, rain . . . there is another motive which is

just as important: the preservation of the group's culture in its pristine form. In this respect the *nagual* plays the same role as was assigned to the Holy Tribunal of the Inquisition during the colony with respect to the dominant Spanish group, that is: to watch over the purity of the customs, to support the hereditary modes of life, and to maintain the integrity of the religious, aesthetic, and economic ideas of the community. (Aguirre Beltrán 1963: 102)

As Aguirre Beltrán (1963: 302) indicated, the characteristics of the nahualli in the colonial period[2] emphasized the worship of Tlaloc and his complex over other gods. This represents a narrowing of nahualli activities, which had a wider scope in the pre-Columbian era when they included divination, astrology, and an association with other deities. The focus on Tlaloc increased the importance of iztauhyatl and yauhtli in generalized curing, amulets, and divination because of the diminished range of deities associated with other plants.

Native healers (*curanderos*) often disguised traditional practices with a gloss of Christianity to avoid persecution by the Inquisition. Thus, the use of hallucinogens, associated like yauhtli with gods, was hidden by assigning them names taken from Christian hagiology; for example, *peyotl* became "Mary's rose," *ololiuqui* became "our Lord," *atlinan* became "Our Lady of the Waters," and yauhtli became "St. Mary's herb." Old remedies and incantations could thereby be used with less danger of detection and denunciation.

Such subterfuge did not fool the early missionaries, whose eye for "idolatry" was sharp. In their accounts of Aztec practices and beliefs, written to help newly arrived priests detect idolatry in all its guises, Ruiz de Alarcón (1982) and de la Serna (1953) gave many examples of syncretism. For example, Ruiz de Alarcón described a cure for a swollen head that involved tobacco and *chalalatli* root followed by a chanted spell:

> After this spell has been said, she blows on the head four times like the curers of Castile; here it is noted firstly, how much the Devil tries to imitate the ceremonies of the Church and secondly, how full of superstition the number four is among the Indians, which alludes to their pagan tradition of the fable of the sun, or is perhaps because the Devil mimics Holy Scripture in that the number four is so generally contained in it, or because through his presumption he adds one to the number

three (which is so full of mystery for Christians). (Ruiz de Alarcón 1982: 232–233)

In a series of explanations of ritual words used in curing
ceremonies, de la Serna wrote:

> they usually mix another two herbs together—*atlinan,* which is
> the water herb, and *iautli,* which is the yerbaniz [African day
> flower], they say that the green spirit and the brown spirit also
> help. To finish, in order to hide their spell, and to make it au-
> thoritative they end with the words—in the name of the Father,
> the Son, and the Holy Ghost. Thus, they mix, as I stated above,
> divine things and Church ceremonies with their idolatries and
> superstitions. (de la Serna 1953: 88)

The records of the Inquisition are a good source of information concerning the "superstitious and pagan" use of herbs. Very
often, the reports were not sufficiently detailed, mentioning herbs
but not giving the names of the specific plants. When specific
names were mentioned, it seems that the principal use of
iztauhyatl was in ritual cleansings (limpias), as described in table
8.3. Aguirre Beltrán (1963: 302) mentioned five Inquisition trials between 1692 and 1775 in which iztauhyatl was so used. Serna (1953:
101) cited one example of iztauhyatl as a cure, Quezada (1975b: 68)
five cases, and Ruiz de Alarcón (1953: 135) another, in which the
nahualli seemed to be using the plant in curing rituals because of
its association with the deity and also by invoking its magical
properties. Syncretism also involved the use of Christian saints'
names for the plants as well as the incorporation of Christian
prayers in the rituals.

Simultaneously, the Spanish settlers began to incorporate native drugs into "academic" colonial medicine because they fit previous conceptions of ailments as caused by "wet" or "cold" and the
prevailing Hippocratic-Galenic doctrine of balancing humors and
qualities. In fact, the ailments being treated were pre-Columbian
in nature. Comas (1954) applied in term "reverse acculturation" to
the adoption by New world physicians, trained in classical European theory, of New World remedies. From the beginning, Europeans were greatly interested in the medicinal products of the
New World; a book by Monardes describing the medicinal plants
of the Americas was published in Spain in 1569, 1571, and 1574,
translated into English by 1577, and rapidly achieved widespread
popularity. Additional evidence of this interest, described in chap-

ter 1, was the funding by King Philip II of Francisco Hernández, his personal physician, for a long expedition to explore the fauna and flora of Mexico.

The process of syncretism can be seen in table 8.4, which lists applications by Europeans in the early (16th and 17th century) colonial period of yauhtli and iztauhyatl remedies, as incorporated into the earliest medical works published in Mexico. The majority of uses were identical with those of pre-Columbian medicine, but the rationale behind them was the Galenic notion of the opposition of "hot" and "cold" qualities rather than the Aztec beliefs. Somolinos d'Ardois, speaking of medicine in New Spain, stated that

> [the patterns] continue to be Galen and Hippocrates. The structure and the doctrine are imported, but in reality the elements employed are American. They are the same ones used by Aztec doctors, infiltrated into the old science and applied according to the classic norm. (Comas 1971)

Reverse acculturation has misled some contemporary observers to claim that the "hot/cold" principle of disease causation in modern folk medicine is due solely to Spanish influence and did not exist in pre-Columbian medical theory (Foster 1953, 1978a,b, 1986, 1987). This topic will be dealt with more fully later in this chapter, but table 8.4 clearly demonstrates that the colonial pharmacopoeia included Aztec remedies for Aztec ailments (eight of the eleven listed). This seems to have been a general pattern, as will be evident for other folk ailments such as fallen fontanelle (*caída de mollera*), evil eye (*mal de ojo*), and fright sickness (*susto*). That is, in the course of medical syncretism during colonial times the etiological explanation of the Indians was lost (or removed), but the character of the illness and its cure remained identical to the native conception rationalized to fit Hippocratic-Galenic principles. Tables 8.5, 8.6, and 8.7 show that this was the case for the uses of yauhtli and iztauhyatl. Although Tlaloc is no longer named as the cause of ailments treated with these herbs in modern folk medicine, these ailments are the ones recognized by the Aztecs not added to the list by the Spanish after the Conquest.

Colonials and native Mesoamericans also began to adopt two herbs of European origin—rue (*Ruta graveolens*) and rosemary (*Rosmarinus officinalis*)—as generic substitutes for yauhtli and iztauhyatl. The native species were considered "hot" and used

Table 8.4. *European Colonial Uses of Iztauhyatl and Yauhtli*

AILMENT	IZTAUHYATL	YAUHTLI	RUE	ROSEMARY	AZTEC USE
Epilepsy	X	—	—	—	yes
Edema	X	X	—	—	yes
Obstructed kidney	X	X	—	—	yes
Stomach, vomiting	X	X	—	—	yes
Colic	X	—	—	—	yes
Madness, craziness	X	X	—	X	yes
Pains due to cold	X	—	—	—	yes
Intermittent fever	—	X	—	—	yes
Worms	X	—	—	—	no
Provoke menses	X	X	X	—	no
Against witchcraft	—	X	X	—	?

Source: Ortiz de Montellano 1986. Reprinted with permission from R. P. Steiner, ed., *Folk Medicine: Art and the Science.* Copyright © 1986 by the American Chemical Society.

Table 8.5. Remedies in Esteyneffer for Aztec Ailments Cured by Iztauhyatl and Yauhtli

AILMENT	IZTAUHYATL	YAUHTLI	ROSEMARY	RUE	STILL USED IN MEXICO
Intermittent fever	X	X	X	X	X
Perlesia, paralysis	X	X	—	—	X
Gout, rheumatism	X	—	X	—	X
Cold disease	X	X	X	—	X
Caused by fear	X	X	—	X	X
Empacho	X	X	—	—	X
Against sorcery	—	X	X	—	X
Lightning burns	—	X	—	—	X
Colic	—	X	—	—	—
Epilepsy	X	X	X	—	—
Dropsy	X	X	X	—	—
Expel phlegm	X	X	—	—	—
Diuretic	—	—	—	X	—
Madness, melancholia	X	—	—	—	—
Fevers	—	X	—	—	—

Source: Ortiz de Montellano 1986. Reprinted with permission from R. P. Steiner, ed., *Folk Medicine: Art and the Science*. Copyright © 1986 by the American Chemical Society.

Table 8.6. Non-Aztec Uses of Iztauhyatl, Rue, and Rosemary in Esteyneffer

AILMENTS	IZTAUHYATL	RUE	ROSEMARY	STILL USED IN MEXICO, SOUTHWEST
Deafness, pituita	X	X	X	X
Diarrhea, dysentery	X	—	—	X
Expel worms-ear	X	X	—	—
Liver obstruction	X	—	X	—
Wound treatment	—	X	X	—
Pujos	X	X	—	—
Hard breasts	X	X	—	—
Flegmon, swelling	X	X	—	—
Rabies	X	X	X	—
Scurvy	X	—	X	—
Prevent apoplexy	—	X	X	—
Blindness	—	X	X	—
Pain in womb from vinegar	—	X	X	—
Sciatica	—	X	X	—
Syphilis	—	X	X	—
Stop vomiting	X	—	—	—
Genital inflammation	X	—	—	—
Bad breath	—	X	—	—
Obstructed spleen	X	—	—	—

Source: Ortiz de Montellano 1986. Reprinted with permission from R. P. Steiner, ed., *Folk Medicine: Art and the Science.* Copyright © 1986 by the American Chemical Society.

Table 8.7. Modern Folk Uses of Iztauhyatl and Yauhtli

DISEASE	IZTAUHYATL	YAUHTLI	OTHER "HOT" HERBS
Susto (fright)	X	X	rue, rosemary
Lightning, fright	—	X	
Witches	X	—	
Fears of pregnant women	—	X	
Aire de cuevas (cave air)	—	X	rue, rosemary
Yeyecacuatsihuiztli = ehecacoacihuiztli Rheumatism, paralysis	X	—	rue
Malaria (atonahuiztli)	—	X	
Cough	—	X	
Colic, empacho	X	X	
Stomach ailments	X	X	
Diarrhea	X	—	

Source: Ortiz de Montellano 1986. Reprinted with permission from R. P. Steiner, ed., *Folk Medicine: Art and the Science.* Copyright © 1986 by the American Chemical Society.

against "cold" diseases by the Aztecs. The European species, classified as "hot" in the Galenic system, were logical substitutes. Evidence for such gradual substitution can be seen in the increasingly overlapping use of rue/rosemary and yauhtli/iztauhyatl in colonial medicine (tables 8.4, 8.5, 8.6). The reason for the substitution may have been either a desire to disguise a ritual connection to Tlaloc or the unavailability of the native species as folk medicine spread to ecosystems such as the northern frontier where the flora was different.

Kay (1977a) has argued that the similarity of remedies used in the folk medicine of various groups in the Southwest must be due to a common source of transmission. She proposed Juan de Esteyneffer's *Florilegio Medicinal*, first published in 1712 in Mexico and widely reprinted thereafter, as the source. The implication, as in Foster's work, is that modern folk medicine, including the "hot/cold" dichotomy, is based on European Hippocratic-Galenic theory because Esteyneffer embraced it as the orthodoxy of his time. Table 8.5 lists the ailments in Esteyneffer for which yauhtli and iztauhyatl were prescribed and which corresponded to pre-Columbian practice. The increased parallelism between Aztec and European herbs can be seen in a comparison between table 8.5 and the earlier colonial usage in table 8.4.

Table 8.6 lists Esteyneffer's prescriptions of yauhtli and iztauhyatl that had no Aztec precedent. Presumably these reflect the logical application of orthodox Galenic theory to the qualities of these plants for extended purposes.

If Foster and Kay were correct, and modern Mexican folk medicine is primarily European in origin, then the primarily European rather than pre-Columbian uses of yauhtli and iztauhyatl would be expected to predominate in modern folk medicine. The exact opposite is the case, as can be seen by comparing table 8.7 with tables 8.4 and 8.5. Of the eleven surviving uses of these herbs in the twentieth century, nine were also used in pre-Columbian medicine, and only two were European introductions. This difference is statistically significant with a chi-square value of 4.45 corresponding to 0.04, that is, a 4-percent probability that the results were due to chance.

= Modern Survivals

Many folk medicinal and ritual ceremonies today cannot be truly understood without taking into account the special relationship between the god Tlaloc and certain illnesses and their cures. Even though the uses of yauhtli and iztauhyatl may now be rationalized in terms of the "hot-cold" system of disease as a consequence of colonial syncretism, the uses are predominantly Aztec. Madsen's (1955, 1969) studies of Tecospa, an Aztec-speaking town near Mexico City, provided an example. Tecospans believed in the existence of *enanitos*, dwarfs who are water spirits, live in caves, send the rain (fig. 8.1) and are responsible for a number of illnesses. When enanitos are offended, they send an ailment called *aire de cuevas* ("cave air"), cured by rubbing the sick person with rue or a bundle of "hot" herbs including rosemary and yauhtli (Madsen 1955: 137). Rain dwarfs are really the tlaloque and tepictoton described in table 8.1 (also see fig. 2.5), and illnesses caused by them are still being cured by yauhtli. Note the use of rue and rosemary in their generic capacities as "hot" herbs. Similarly, In-

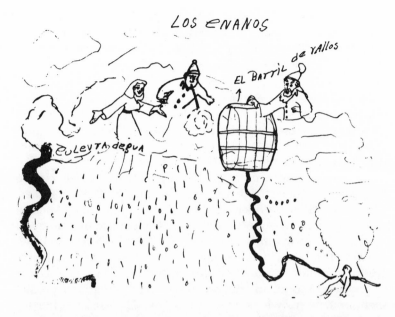

Figure 8.1. Rain dwarfs equivalent to Tlaloque painted by a seven-year-old informant in the Aztec village of Tecospa in the 1950s. The influence of Hollywood is evident. The dwarfs, like their fifteenth century predecessors, afflicted people with "cold" illnesses. (Madsen 1960. *The Virgin's Children*, by permission of William Madsen)

gham's (1986: 116) informants described dwarfs called *yeyecame* ("airs") that dwell in anthills and another spirit, called Rainbow, that dwells in caves and causes susto, dropsy, and fever with chills, diseases related to Tlaloc.

The rain dwarfs in Tecospa also send a very "cold" disease, *yeyecacuatsihuiztli*, with symptoms (paralysis, twisted mouth, pain in the joints, and palsy) resembling rheumatism (Madsen 1955: 137; 1969: 630). The name of the disease is a corruption of *ehecacoacihuiztli*—"the coacihuiztli caused by air." It is cured by ritual cleansing (limpia), and

> mentholatum (hot) is rubbed over the patient's body and he stands naked over a pot in which boils teposan (hot), estafiate (hot), ruda (hot), a glass of vinegar (temperate), and a jigger of

alcohol (temperate). Both the patient and pot are covered with a
blanket where they remain until the liquid has evaporated.
Then the patient retires well covered and sweats the coldness
of the disease out of his body. (Madsen 1955: 137)

Here we see the use of iztauhyatl along with the generic rue and
rosemary to cure a "cold" disease, but, as seen in tables 8.2a and
8.2b both the disease coacihuiztli and the remedy are pre-
Columbian, when the presumed physical mechanism of the cure
was the removal of "phlegms" by induced perspiration.

A group of apparently unrelated rituals and beliefs can be
linked by the association between yauhtli and the Tlaloc complex.
September 28–29 is widely celebrated as the Feast of St. Michael
who is believed to fight with the devil that night and prevent harm
to crops by strong winds (Díaz 1975: 163; 1976: 109; Bock 1980: 142;
Ingham 1986: 142). On the previous day, flowers of yauhtli are gathered, made into crosses (often blessed by the priest), and placed at
the corners of cornfields, doors of houses, and granaries to protect
them from strong winds (Lewis 1970: 140, 461; Diaz 1975: 163; Ingham 1986: 59). They are even used in commercial agricultural
fields and on passenger buses (Bye 1986, pers. com.). In Santiago, a
small valley within the Municipio of Tepoztlan, dried flowers of
yauhtli are thrown into a fire because their smoke drives away
clouds that threaten hail and possible damage to crops and
houses (Bock 1980: 142).

The syncretic sequence was clear in Tepoztlan. It was believed that the winds damaged crops on September 29 because
dancers representing El Tepozteco, the Catholic patron saint of
the village, had not performed properly on his feast day, September 8 (Lewis 1970: 150, 461). El Tepozteco is just the Spanish version
of Tepoztecatl, the original Aztec *capulteotl* ("deity of the town").
Tepoztecatl was one of the "400 Rabbits," gods of alcohol and fertility and members of the Tlaloc complex. The dance to El Tepozteco
in front of the church can be interpreted as an age-old ritual to
ensure rain and good crops. Over time the functions of the Wind
God, Ehecatl, and Tlaloc fused. As mentioned above, during the
colonial period the roles of various gods tended to be channeled
into Tlaloc alone as the pantheon was simplified due to the absence of trained priests. Aztec religion had linked Tlaloc with
Ehecatl because winds usually presaged rain during the rainy season (Sahagún 1956: vol. 2, 264; Seler 1900–1901:42–43). Thus,
Tepozteco's choice of wind as a method of punishment can be
traced to the persistence of pre-Columbian patterns of belief, and

yauhtli, Tlaloc's herb, is the logical amulet to ward off harm. Further syncretism is indicated by the circumstances that September 8 is the Feast of the Nativity, and the townspeople believe that El Tepozteco was the son of the Virgin Mary. As late as 1930, the village priest would bless the hill where El Tepozteco lived so that he would not ruin the crops (Lewis 1970: 253–256; López Austin 1973: 138).

Guilds of healers exist called *graniceros* ("hail people"). They have strong shamanic attributes (such as ingestion of hallucinogens and assignment to the vocation of healing by being narrowly missed by a lightning bolt) and are reputed to attract rain, prevent hail, protect crops, and cure people (Cook de Leonard 1966; Ingham 1986; 157). Viesca Treviño (1981, pers. com.) described a pilgrimage graniceros undertake yearly on May 3–5 (the Day of the Holy Cross) to a cave on the mountain *Iztaccihuatl* to renew their powers. Firecrackers are used to simulate lightning and so attract beneficent beings. The cave is swept out with a broom made of yauhtli to remove two "evil winds" called the "Bull" and the "Snake," spirits that can deprive the graniceros of all their powers. Inside the cave, a blue-green gross (the traditional color for Tlaloc in Aztec symbolism) is worshipped. In Tlacayapan the graniceros (whose patron saint there is St. Michael) also make a power-renewing pilgrimage to a cave on May 3, the beginning of the rainy season (Ingham 1986: 158). Their ceremony also involves the use of yauhtli. Syncretism involving Tlaloc was also described in connection with ceremonies at the Sanctuary of Chalma where Hobgood (1971) found many green crosses in nearby caves. The crosses were brought to town, washed, painted with a new coat of green paint, erected, and decorated with paper flowers and honored with dances and offerings.

Syncretism can also be seen in Cortés's (1979) account of the curing of *mal aire* ("bad air") disease. The cure involved a ritual cleansing (limpia) in which both Catholic and the "Spirit of Lightning" were invoked. Here we see the blending of both Tlaloc (lightning) and Ehecatl (wind) with Catholicism.

Syncretism in Latin American Folk Medicine

In a classic paper, Foster (1953) proposed that a number of elements in Latin American folk medicine, such as the "hot-cold" classification of food and disease and certain culture-bound syndromes, were derived exclusively from the Spanish tradition with no native contribution. The culture-bound syndromes he identified as purely Spanish were: *caída de mollera* ("fallen fontanelle") and *mal ojo* ("evil eye"); a third common folk illness, *susto* ("fright"), was presumed to be of native origin. Foster's judgments have been widely accepted and for a long time were not disputed. Although scholars like Isabel Kelly and Diana Ryesky expressed some reservations, the principal opponent of Foster's conclusions was Alfredo López Austin. López Austin's views, however, have been published only in Spanish until recently and thus not widely known in the United States.

We will discuss Foster's thesis as expressed in his original paper and in subsequent articles, starting with the "hot-cold" classification system, the subject of greatest controversy. The question of the origin of folk diseases will be dealt with on a case-by-case basis. The discussion will not cover all of Latin America since that is beyond the scope of this book.

Foster (1953) observed that both Indian and mestizo populations in Latin America have a traditional classification of foods, illnesses, medicines, and people as "hot" or "cold." Humans must achieve equilibrium to be healthy. A hot-cold imbalance must be redressed by the ingestion of contrary elements. Foster postulated that the wide dispersion of this belief in former Spanish colonies indicated its roots in the Hippocratic-Galenic system prevalent at the time of the Conquest. This system recognized four qualities—dry, wet, hot, and cold—associated with the four elements—earth, fire, water, and air—that were the basis of all matter. In the New World, however, the fourfold classification has become twofold; the wet-dry dimension was lost. Foster (1978a) later attributed this loss to the difficulty of maintaining and remembering too many variables in nonliterate civilizations.

López Austin (1988: vol. 2, 281) held that Foster greatly underestimated the information-retaining ability of societies that depend on oral transmission. Studies of folk botanical taxonomies (Berlin, Breedlove, and Raven: 1974) showed that preliterate people can retain and transmit vast amounts of information. It is also difficult to explain why the wet-dry classification was the one that was universally lost. Foster (1986) argued that "hot-cold" was retained because these qualities are more "salient." However, since "hot" and "cold" are often metaphorical rather than actual physical properties, relative salience is far from obvious. A much more likely explanation, at least for Mesoamerica, was proposed by Madsen (1955: 138):

> the Hippocratic system was intimately compatible with the ancient Aztecan concept of the Universe ordered on a system of balancing opposites. This compatibility between the pre-existing Indian configuration and the introduced European complex undoubtedly favored the acceptance of the Hippocratic system by the Indians. The European hot-cold complex was meaningful to the Indians because it could be fitted into the familiar Aztecan complex of eternal war between hot and cold. . . .

> In the process of assimilating the Hippocratic system the Indians were eclectic. They did not attempt to imitate the Spaniards by taking over the new complex intact but rejected parts of it such as the wet-dry classification which evidently was not referable to indigenous belief. In Tecospa today, wetness is significant in the hot-cold classification only insofar as it produces the quality of coldness, and dryness is significant only because it indicates the lack of wetness and the associated quality of coldness.

This fits well the example of Tlaloc's herbs discussed earlier, for which wetness and cold were often joined in describing ailments to be treated, while dryness was not regarded as a distinct category. Also, the etiology and cure sometimes clearly made this distinction. Dropsy, for example, was due to an excess of water and was cured by a diuretic not by a "dry" remedy, and coacihuiztli, a "cold" disease, was cured by a "hot" remedy.

Foster (1986) raised the question: Aztec ideology may have been receptive to and thus fostered syncretism, but why does the classification system exist elsewhere in Latin American in the absence of Aztec ideology? The question is not relevant. What must be kept in mind are the two opposing paradigms and the levels of proof they involved. The prevalent view that Latin American humoral folk medicine derived solely from European tradition is absolutist, denying any syncretism with native traditions. As Foster stated:

> I believe that the humoral theory, beliefs, and practices found in contemporary Latin America (and Christian Philippine) popular medicine are directly descended from classical Hippocratic-Galenic humoral pathology, essentially unaffected by Aztec

and other indigenous Hot-Cold theories. Indigenous New World herbal and other remedies, of course, have been incorporated into the humoral framework and continue to be widely used. But this incorporation is unrelated to pre-Conquest theories. (Foster 1986)

The burden of proof is therefore incumbent on proponents of the absolutist theory.

The contending paradigm claims only that modern Latin American humoral folk medicine is the result of syncretism between compatible elements of the native tradition and European beliefs. To assess its validity, it is necessary to investigate folk illnesses and beliefs case-by-case without the a priori assumption that beliefs in all countries are fundamentally similar. The burden of proof is lighter; all that must be demonstrated is the contribution of syncretism to a particular folk illness or "hot-cold" classification in a particular country, not necessarily in all countries or all cases. The Foster paradigm excludes native contributions in all cases everywhere. Widespread belief does not prove Spanish origin since susto, an acknowledged non-European folk illness, is also widely dispersed in Latin America.

Foster (1986) also claimed that the existence of a "hot-cold" system in places as diverse as Haiti, Puerto Rico, and the Philippines certifies its a Spanish origin. However, the history of these three localities was so different from that of the rest of Latin America that their possession of a dual humoral system has little explanatory value, and different reasons may be needed to explain the decay, if such is the case, of an introduced Hippocratic-Galenic system. The Indians of Puerto Rico were annihilated shortly after colonization, so Foster's reference to a nonliterate native population to explain the loss of wet-dry classification is not applicable there. In the case of the Philippines. López Austin (1988: vol. 1, 280) pointed out that the Chinese, who also had a fourfold humoral system, traded with the islands nine centuries before the arrival of Europeans and are a more likely source than the Spaniards.

López Austin's (1971a: 21–41; 1986: 303–317) main arguments in support of syncretism were: 1) The hot-cold polarity is not limited to the areas of health, illness, and medicine but extends to the whole cosmos. 2) The system was mentioned in very early colonial sources. 3) Many concepts in folk medicine and Aztec beliefs are not explicable by or contradict classical European theory. These arguments and Foster's responses will be expanded on below.

Very early Colonial documents refer to beliefs regarding "hot-cold" among the natives. Hot and cold medicines and illnesses were mentioned in Sahagún's works, in the *Badianus Codex*, and the works of Francisco Hernández. Redfield and Villa Rojas (1964: 372) cited Martin de Palomar and Gaspar Antonio Xiu who wrote in 1579 that the native Maya could explain the properties of medicinal plants only in terms of their "hot" or "cold" qualities. Hernández (1959: vol. 1, 344), referring to plants generically called *totoncaxihuitl* ("hot herb"), stated that "they call *totoncaxihuitl* sometimes those plants which fight against heat but more frequently those which are hot in nature." As discussed in chapter 1, Hernández often classified plants according to his own classical criteria. Of the fifteen plants in that section, called "hot" by the Aztecs, he classified five as "hot" and ten as "cold," which implies both that the Aztecs made such classifications and that their criteria differed significantly from classical Hippocratic-Galenic theory.

Foster (1978a,b; 1986) rejected such examples on the basis that the Colonial sources were contaminated by European influence. He said the Hispanicized Indian authors had been trained using European books, and references to Pliny in the *Badianus Codex* demonstrated the European bias.

The whole concept of tonalli, described in chapter 2, demonstrates the Aztec belief that heat and cold affect health and physiology was a fundamental component of their worldview and had wide ramifications into areas beyond health and illness. Some Aztec beliefs reported in various sources can be explained only in terms of tonalli and are completely alien to European culture or unexplainable in European terms. López Austin (1988: vol. 1, 260–262) showed that both folk beliefs today and Aztec beliefs related heat to tiredness and cold to rest and relaxation. The relationships are evident in etymology. To rest and "to cool hot things" were, respectively, *cecelia (nino)* and *cecelia (nite)* in Nahuatl; *cehuia (nino)* and *cehuia (nite)* were, respectively, "to rest" and "to cool hot things down." *Tlacehuilli* can be translated as either "cooled thing" or "a thing relieved from work." *Moceltiqui*, literally "he that cools down," means "he who rests, recreates," and *tonalcehuia (nino)*, literally "to cool the tonalli," means "to rest someone who walks."

To avoid harm the Aztecs would cool their tonalli by resting and by drinking pulque before doing heavy work (López Austin 1988: vol. 1, 261). That pulque was classified as "cold" even though it contained alcohol was shown by the belief that pregnant women, who had an excess of heat, should not serve pulque because they would rob it of its power to intoxicate (López Austin 1969: 139). This concept was alien to the Spanish and obviously not introduced by them, as the following passage by Serna (1953: 285) showed:

> in all the rest of the world men rest after work, in this nation it is the opposite, the natives rest before working, a custom that is depraved and unreasonable. Since they are all subject to work, either in their own cornfields, or in compulsory personal service, or by long trips, the Devil (or an old evil custom) has persuaded them that, if they will get very drunk and very often before starting these tasks, they will gain strength and the ability to perform the pending work. The call this action *nocehuiliztli* which means "my rest, to gather strength for work" in their language.

Exercise thus raised the temperature of the tonalli but simultaneously cooled other parts of the body, particularly the belly and the feet. It was then necessary to rub the affected parts with a "hot" remedy such as *picietl* (tobacco). The "hot" nature of tobacco was also implied in its use against "cold" diseases such as coacihuiztli, caused by Tlaloc, and as a remedy for physical cold by rubbing it against the belly (Alva Ixtlilxochitl 1975: vol. 2, 74). In chapter 7 the plant *tlacopatli* was described as having a special affinity for tonalli. Hernández (1975: vol. 2, 127, 130–131) described various species of the plant as "hot" and said it was used to revive and strengthen those who were tired. It was used against a variety of "cold" diseases such as intermittent fevers, for example, *atonahuiztli* ("aquatic fevers"), associated with Tlaloc. In this case, the plant's "hot" classification was clearly also Aztec because of its special affinity to attract and hold tonalli.

López Austin's point that the hot-cold duality pervades the Aztec worldview is supported by material already discussed in previous chapters. How could European intrusion be responsible for this duality when it was already an integral part of a system that related otherwise disparate beliefs clearly not European? For example, the concept of tonalli (heat in the body) explained a vast number of different Aztec beliefs and practices, many completely foreign to European experience, concerning such things as *chipil* illness, the "cold" nature of twins and their effect on sweatbaths and dyes, the right of nobility to rule, the need to burn a fire continuously in a newborn baby's room, drinking pulque before work, fright illness, human growth and development, and the relationship between fate and the calendar, to cite just a few. Similarly, yauhtli and iztauhyatl were classified as "hot," but this was not merely an isolated fact. It was part of a complex that linked the two plants to *capulteotl* ("local patron gods"), gout, paralysis, fevers, malaria, being narrowly missed by lightning, epilepsy, madness, leprosy, the Rain God complex, caves, tlaloque, tepictoton, etc. The "hot-cold" duality was clearly an integral part of Aztec culture before the arrival of the Spanish.

Foster (1986) made the further point that the Aztec system, as presented by Sahagún, treated the nature of listed remedies asymmetrically:

> among the Aztecs Cooling remedies predominated. While one does not expect absolute symmetry in a humoral system, substantial representation of both Hot and Cold remedies is assumed. What does the evidence show?

> In López Austin's Spanish translations, I count 11 *Frio* or *Fresco* remedies and 10 more that *Enfria* ("Cool"). No remedy is described as *Caliente* or *Calentando* ("Heating"). In the Dibble-Anderson English translation I count 12 remedies described as Cold or Cooling, but not one that is Hot or Heating. An additional 75 or so remedies in the *Florentine Codex* are prescribed for fever, with such expressions (in Spanish) as *calma el dolor, para fiebre, para cuerpo caliente*. With a humoral frame of reference, it can be argued that these remedies must be Cooling, although this is not made explicit.

> In contrast to numerous implicitly Cooling remedies there are very few implicitly Hot ones. The herb *oquichpatli* is an example: "It is required by one whose member has been harmed, and by one whose urine has been stopped. . . . When they have given the medicine, he becomes very hot; the sweat exudes (Sahagún 1950–69:book 11, 185). This suggests a Hot remedy, . . . López Austin also lists *oquichpatli* as well as seven additional remedies for "chills and fever" and "aquatic fever," which he equates. Even if one accepts these eight examples as implicit evidence of Heating remedies, the fact remains that the humoral evidence in Sahagún is incredibly lopsided, quite out of line with any contemporary system.

These conclusions can be reached only by ignoring completely all Aztec etiological concepts. For example, Foster claimed that 75 remedies were described for fever and that these must implicitly be cooling. As shown previously in Ortiz de Montellano

(1979b) and in previous chapters, the Aztecs believed that fevers were caused by hot phlegms in the chest whose cure was expulsion by an emetic, a purgative, a diuretic, or a diaphoretic (clearly not a cooling remedy!). Many of Hernández's disagreements with native classifications were due to his insistence, similar to Foster's, that fevers should be treated with cooling herbs while the Aztecs used herbs he considered "hot." He stated (Hernández 1959: vol. 1. 273) that "they believe that heat is cured by heat," and (1959: vol. 1, 132) "[tlacopatli hot and dry] . . . is taken to produce sweating and in this manner (as the Indian doctors themselves assure) fevers are relieved." In fact, as discussed earlier, a number of these remedies contained diaphoretic chemicals and the sweating they induced was consistent with the Aztecs' etiological beliefs but did not fit the European humoral perspective.

Foster erred in saying that Sahagún does not explicitly refer to a hot remedy. Sahagún (1950–1969: book 10, 90) states:

> The essence of *axin* [is] hot; they say [it is] like fire. The traveler anoints himself with it in order that the cold will not oppress him exceedingly [similar to the use of tlacopatli for cold feet due to overheated tonalli]. Where the sickness named the gout [coacihuiztli] occurs, wherever it is, [axin] is spread on. It soothes [the gout]. In order that frost will not injure the lips, an *axin* unguent is applied; it is spread on the lips. . . . This is not used alone as a cure for gout; it is mixed with the herb called *colotzitzicaztli.*

Hernández (1959: vol. 1, 255) confirmed this classification: "[*temecatl* mixed with the grease of worms called *axin* . . . they say it cures pains of cold origin admirably." Thus, the cure for gout (coacihuiztli), one of Tlaloc's diseases, was clearly "hot" medicine, and, implicitly, other gout remedies such as tobacco (mentioned above), yauhtli, and iztauhyatl also were "hot." Since Sahagún's informants seldom named explicit humoral qualities, most determinations have to be implicit and based on a knowledge of Aztec etiological concepts. In any case, allegation of a drastic imbalance in the humoral qualities of Aztec remedies is unfounded.

Ironically, the alleged imbalance itself tends to refute a Spanish origin for the "hot-cold" designation in Aztec sources. As Foster (1953: 205) himself stated, classical humoral theory (the supposed source) also was unbalanced but in the opposite direction:

Nevertheless, certain general rules seem to prevail; the most marked is that (following classical theory which believed a preponderance of heat to be the normal state of the healthy body and undue cold as the condition most frequently needing remedy) a majority of medicinal herbs are classified as hot.

The essence of the counterargument to Foster's view is that the "hot-cold" polarity pervaded Aztec thinking, including concepts of health and illness, before the Spanish came. The Aztecs were receptive to European humoral theories that resembled their own, as in the syncretism involving "hot-cold" duality, and rejected the alien concept of "wet-dry." This explanation fits the known facts better than Foster's. The explanation does not necessarily apply to how the rest of Latin America acquired their systems, which requires the same kind of analysis for each country as was done here for Mexico. (For articles arguing against the Foster hypothesis in other areas of Latin America, see Bastien: 1985; Colson and de Armellada 1983.) The evidence for a pre-Conquest "hot-cold" duality in Aztec culture makes Foster's exclusion of all native contributions to the fundamentals of Mesoamerican folk medicine untenable.

Culture-bound Syndromes

Foster (1953) mentioned several folk illnesses, including susto, mal de ojo ("evil eye"), and caída de mollera, claiming an exclusive Spanish origin for the latter two. We shall analyze all three to show their Aztec-European syncretic content, with different contributions from each tradition.

Susto

Susto, a consequence of fright, is ubiquitous in Latin America. Rubel (1964) reported:

> One of the most noteworthy aspects of the *susto* phenomenon is the fact that a basic core of premises and assumptions—symptoms, etiology, and regimens of healing—recur with remarkable constancy among many Hispanic-American groups Indian and non-Indian alike.

The causative factor is a frightening experience that releases the soul from the body. Loss of soul in turn produces the symptoms:

restlessness in sleep, listlessness, loss of appetite, weight loss, loss of energy and strength, depression, introversion, paleness, lethargy, and sometimes fever, diarrhea, and vomiting. Crandon (1983) pointed out that these could be symptoms of a variety of diseases, such as gastroenteritis, amebiasis, hookworm, trichinosis, deficiencies of various B vitamins, and hypoglycemia. Other suggested diagnoses ranged from psychiatric—a culturally appropriate expression of hysterical anxiety (Gillen 1948)—to hypoglycemia (Bolton 1981). Rubel et al. (1984) attributed the symptoms to social stress but showed that susto victims (*asustados*) also suffered higher levels of organic disease. In a prospective study, people with susto subsequently died at a statistically significant higher rate than a matched control group. Although the causes of death varied, stress may well have damaged organ systems to a degree that depended on personality or genetic factors so that a common underlying cause may have led to eventual death from a variety of ailments.

Susto clearly had origins in pre-Columbian native cultures, as Foster conceded (Foster and Anderson 1978: 76), since a belief in multiple souls was decidedly not European. In Mesoamerica, the connection was evident:

> The etiological explanation stems from pre-Hispanic times and has survived to the present with incredible persistence. In the Nahuatl language, the ailment is called *tonalcahualiztli,* a word that means literally "loss of soul—*tonalli.*" . . . Its loss was the result of a severe fright, an emotional shock, that unbalanced the organism and permitted the *tonalli* to leave. Cure consisted in going to collect the soul-*tonalli* wherever it might be found or from whoever might have appropriated it, in order to return it to its legitimate owner. This concept should be kept in mind to explain the treatment by magical rituals that is still practiced and that includes "cleansing" prior to recovery of the soul, which expels from the body all the bad spirits and airs that take advantage of the exit-place of the *tonalli* to lodge themselves in the body (López Austin 1978: 201)

It is interesting to speculate on the widespread incidence of susto, from Chicanos in the United States to Peru and Chile to Carib-speaking Indians on the border of Brazil and Venezuela (Colson and de Armellada 1983). Since it is not of Spanish origin, Foster's hypothesis of a common colonial heritage cannot be invoked. As we will see below in the case of caída de mollera, not all these illnesses are uniformly distributed in Latin America. One possibility is that susto is a relic of Paleolithic shamanism (described in chap. 2). Multiple souls, some of which can leave the body, are integral parts of shamanic religions. If the underlying cultural stratum for all of the New World was shamanism brought over by the Paleolithic hunters who settled the continent, its preservation would explain why a belief in soul loss would span the continent. Evidence for this can be seen in the differences in susto beliefs between Indian populations and mestizo and white groups. All are affected by susto and apply similar cures—predominantly magico/religious rituals involving the use of herbs or other substances in "cleansings" combined with prayer or incantations—but differ in etiological explanations. As in the case of Tlaloc's herbs and caída de mollera, syncretism with European ideas tended to dislodge native etiological explanations among the mestizos and whites while both the disease and the cure were retained. Thus, in Mexico (Signorini 1982), Bolivia (Crandon 1983), Texas (Trotter 1982), and Chile (Simmons 1955) more acculturated people still believe in and treat susto but do not associate it with soul loss. The etiology is limited to shock or fright. Indian populations, however, retained the concept of multiple souls and their possible loss. Some syncretism has taken place in their view of susto, but the predominant influence is clearly Indian rather than European.

MAL DE OJO

The opposite type of syncretism took place with mal de ojo ("evil eye") where the predominant influence is European. For the Aztecs at least, compatibility with previously held beliefs made adoption of European ideas easier for the native population. Belief in the evil eye probably began in antiquity in the Middle East and diffused widely from there (Maloney 1976:xi). Infants and children are most at risk from mal de ojo. The illness is triggered by a look from someone with a "strong vision." In most cases strong vision is involuntary and not controlled by the possessor. Less frequently the power is believed to be voluntarily acquired through a pact with the Devil (Aguirre Beltrán 1963: 26). In Guatemala the power may be inborn or due to a transitory state of "hot" blood due to exercise, pregnancy, menstruation, or drunkenness (Cosminsky 1976: 165). "Hot" blood caused by exercise is also given as a precipitating factor for strong vision in Peru (Simmons 1955). Another variant attributes mal de ojo to someone who casts an envious glance at a child (Rubel 1960).

Symptoms of the illness brought on by mal de ojo are most often a child's constant crying, fever, and diarrhea. Remedies are fairly uniform and magical/religious in nature. The illness is diagnosed by rubbing a hen's egg over the patient's body, often tracing a cross and sometimes accompanied by prayers, then breaking the egg into a glass of water and allowing it to settle. The form it takes indicates the illness and often the source. The cure may involve a second "cleansing" with an egg or the use of herbs; rue in particular is used in several countries. Preventive measures are also fairly uniform and usually involve the color red. Children wear necklaces or bracelets of bright red seeds from the coral tree (*Erythrina corallodendron*), red clothing, or other red items (Cosminsky 1976: 167). Interestingly, red was also the predominant color used in amulets to ward off the evil eye among Jews and in Mediterranean countries, and a sprig of rue was placed over the door in Italy (Moss and Cappannari 1976: 7–8). A Spanish author, Pedro Ciruelo, quoted by Madsen (1967: 129) described beliefs about the evil eye in 1538 contemporaneously with the Spanish conquest:

> Some unhealthy men and women can infect others and endanger their health with their vision and with the breath from their mouths. This most commonly happens to young children . . . for the infection easily penetrates these victims when persons with the evil-eye look at them at close range and talk to them. This sickness is caused not only by the eyes but also by breath from the nose and mouth and by the sweat or odor or vapor exuded by the entire body of the contagious person. Evil-eye sickness should be cured by doctors and by commending oneself to God and the saints, never by consulting witches or other persons who cause evil-eye sickness.

> If evil-eye sickness is not the result of close contact with a person of unhealthy vision, there is reason to suspect that the sick person has been bewitched by malicious witchcraft because persons of unhealthy vision cannot cause evil-eye sickness from afar. Evil-eye sickness caused by witchcraft should be exorcised by a priest and not cured by another witch.

Although the primary source of the mal de ojo syndrome in Mesoamerica was clearly European, a particular pre-Columbian belief made it more readily acceptable. This concerned ihiyotl and the group of illnesses called tlazolmiquiliztli ("garbage disease") discussed previously. Here, too, baneful emanations made children and other weak beings sick. Sexual sinners emitted ihiyotl, which could harm innocent bystanders much the same as the mal de ojo. Nahualli could emit ihiyotl voluntarily and harm others just as, in European belief, those who dealt with the Devil had the power of the evil eye. In the Mexican village of Tlayacapan, studied by Ingham, the similarity between ihiyotl and mal de ojo was manifest. The villagers beliefs in mal de ojo followed the standard pattern: 1) children are the principal victims; 2) the causative agent is an envious glance; 3) diagnosis consists of rubbing the patient with an egg then breaking it into a glass of water and examining the pattern it makes; 4) the use of red amulets such as sea coral to ward off the disease (Ingham 1986: 164–165). Another observation by Ingham indicated a direct descendant of tlazolmiquiliztli:

> People also speak of *aire de basura* (garbage air). It may be given off by men or women who have had sexual relations recently, but is emitted especially by prostitutes or loose women. *Aire de basura* is hot and similar to the offending agent in evil eye sickness. . . . People who are hot with anger also can cause *daño de muina* (harm from anger), a sickness similar to evil eye sickness. Garbage air sickness is treated by some people in the same manner as evil eye sickness. . . .

> Belief in evil eye sickness may be part of the constellations of beliefs about *aires*. *Aire* illness, evil eye sickness, garbage air illness, and harm from anger are all similar. Presumably, in evil eye sickness the adult's eye made hot by passion of sex or envy expels an *aire* or, alternatively, attracts an *aire* and then becomes a vehicle for transmitting it to the child. Eggs are used to diagnose and cure evil eye sickness and the various *aire* ailments. The presence of both evil eye sickness and *aire* sickness is indicated by small "whirlwinds" in the albumen . . . both evil eye sickness and *aires* affect the victim's eye—usually the left— and indeed have a special attraction to eyes, which no doubt accounts for the popularity of the deer's eye [*Thevetia sp.*] as a charm against the evil eye. In support of this interpretation, it may be noted that in medieval Europe evil spirits were thought to enter and leave the body through the eye. (Ingham 1986: 165–166)

CAÍDA DE MOLLERA

Although Foster (1953) claimed that caída de mollera had a pure Spanish origin, and Kay agreed, claiming further that

Esteyneffer's *Florilegio Medicinal* was the means of transmission to the Southwest, the syndrome is actually the product of syncretism with a predominant Aztec influence (Ortiz de Montellano 1987). Caída de mollera is found only in infants. It involves a depression of the fontanelle of the parietal bone of the cranium (the *mollera*). Belief in the culture-bound syndrome is still widespread among Mexican Americans (Graham 1976; Scheper-Hughes and Steward 1983); in Mexico (Madsen 1965; Madsen 1960; Scheffler 1977; Foster 1967: 193; Casillas 1978; Young and Young-Garro 1982); in both rural and urban areas in Guatemala (Knoke 1984: pers. com.; Hurtado Vega 1979; Cosminsky 1980); in Honduras (Kendall, Foote and Martorell 1983); and among descendants of the Pipil in El Salvador (Clement-Brutto n.d.). This distribution is significant in tracing the origin of the syndrome.

Folk etiology attributes caída de mollera to either a fall or the sudden withdrawal of the nipple from a nursing infant, which causes the palate to fall and produce a "sunken" fontanelle. The symptoms of caída—restlessness, crying, and a failure to eat—are believed to result from blockage of the oral passage by the sunken palate. These symptoms correspond physiologically to severe dehydration (Trotter, Garza, and Garza n.d.). A variety of remedies are used in attempts to restore the palate to its normal position. Most commonly the palate is pushed up with the finger, or the child is held upside down and shaken (Harwood 1981: 307; Kay 1977b; Hurtado Vega 1979; Holland 1978: 103; Rubel 1960). Other cures include poultices (Martínez and Martin 1966; Kay 1977b: 150; Rubel 1960) and suction of the fontanelle either directly or by holding the child upside down while dipping its head in and out of water (Harwood 1981: 307; Martínez and Martin 1966; Kay 1977b: 150; Rubel 1960).

Trotter (1985) conducted one of the largest surveys of Mexican-American beliefs about culture-bound syndromes, collecting data from a random sample of 1,900 individuals in 35 public health clinics in the Southwest and from a purposive sample of 80 individuals in the Rio Grande Valley who had treated cases of caída de mollera. The survey confirmed the existence of a widespread belief in caída de mollera, with local variations ranging from a low of 2.7 percent in Presidio, Texas, to a high of 69.9 percent in Tucson, Arizona. Table 8.8 lists the various remedies in the order of the frequency of reported use. Pushing up on the palate and/or shaking the child upside down were by far the most frequent, while such remedies as an egg poultice or dipping the child's head into water were used by only a small percentage of practitioners.

Table 8.8. Treatments for Caída de Mollera

TREATMENT	THREE-STATE SURVEY (N = 604) % of N	RIO GRANDE (N = 175) % of N
Pushing up the palate	40.7	30.2
Turn upside down	31.0	27.4
Soap in fontanelle	9.4	16.0
Suck fontanelle	6.8	5.7
Egg in fontanelle	3.6	5.1
Dip in water, upside down	1.3	2.9
Various others	7.2	12.7
Total	100.0	100.0

Source: Ortiz de Montellano 1987: 385. Copyright © 1987 by Duke University Press.

Foster (1953) claimed that caída derived from a group of illnesses in Spanish culture thought to be caused by displacement of body organs from their normal position and treated by repositioning the organ. He identified Spanish precursors as *espiñela* ("bones in the pit of the stomach"), *paletilla* ("bones between the shoulder blades"), and *calleiro* ("fallen stomach"); the first two could be caused by violent exercise or coughing. Caída de mollera was and is not present in Spain, but, since folk etiology now attributes the problem to a physical displacement of the palate, Foster felt that the Spanish organ-displacement ailments, by analogy, were the model for the New World syndrome. If, as Foster proposed, caída de mollera is exclusively Spanish in origin, then one would expect logically to find evidence for the following: 1) reference to the syndrome in medical texts available to the Spanish during the sixteenth century; 2) the presence today of caída de mollera, or an analogous syndrome, in the ethnomedical tradition of Spain; 3) a fairly even current distribution of the trait throughout contemporary Latin America, for example, as with mal de ojo; and 4) predominance of the Spanish treatment in contemporary Latin America. Such evidence is not found.

Foster never cited a direct European counterpart to caída de mollera or references to it anywhere in the Old World. A number of medical historians have stated that fallen fontanelle is not a known European folk illness and does not occur in any texts familiar to them (Moss 1983, pers. com.; Majno 1983, pers. com.; Risse 1984, pers. com.). Kay 1979; 1986, pers. com.) claimed that Pliny mentioned fallen fontanelle: "If the uvula be fallen, it will up again, if the patient suffer another to bite the hair in the crown of the head, and so pull him plumb from the ground" (Turner 1964: 274–275). The condition described by Pliny was not a valid precursor because : 1) the uvula is not the palate, the organ that falls according to the etiology of caída; 2) no indication was given that the crown of the head was sunken; and 3) the passage was not in a section dealing with children's diseases so it could also apply to adults. Sunken fontanelle can occur only in small children, whose bony sutures have not yet closed.

More important questions for our purpose are whether Esteyneffer had access to or was influenced by Pliny and, more generally, what sources influenced his book. It is not valid argument to pluck citations here and there from various authors in antiquity to demonstrate that a particular concept prevailed at an earlier date or influenced a particular author. One must show that the belief was contemporaneous and generally accepted or that the author in question had access to particular works (by citations or internal evidence). There is no evidence that Esteyneffer used Pliny or Soranus (discussed later) in preparing his work. As we will see, the treatment for relaxed uvula in Pliny does not resemble in the least Esteyneffer's treatment for fallen fontanelle. Carmen Anzures, editor of the new standard edition of Esteyneffer's work, lists the sources Esteyneffer used (Esteyneffer 1978: vol. 1, 34, 67) as the works of Hippocrates, Galen, Boerhaave, and Sydenham. A search of the index of the complete works of Hippocrates (Littre 1846) revealed no reference to fallen fontanelle or its treatment, nor does it appear in the works of Sydenham (Latham 1848). Boerhaave (1755: 68) mentioned skull depression only in the context of treating wounds and fractures. Kay herself (1986, pers. com.) pointed out that organ displacement was not a part of Galenic medicine.

Kay (1986, pers. com.) argued that the term "fallen fontanelle" is not found in the classical text because it was known under a different name and based on a different etiology. She claimed that it was called *siriasis* or *sitibundum* and related to a burning fever or inflammation of the brain. Soranus's work is a standard source:

> Siriasis, as Demetrius states in his book on "Semiotics," is nothing but a burning fever. According to some people, however, it is

an inflammation of the parts around the brain and meninges so that as a consequence the bregma [front part of the head] and the eyes sink in; at the same time there is a paleness and dryness of the body and anorexia. When siriasis occurs, one should do everything as in a case of inflammation. Now some people say that the disease is named seiriasis after the star because of the heat, but according to others it is called after a sunken bregma because among farmers *seiries* is the name of a hollow object in which they throw and keep their seed. These patients too benefit by the yolk of an egg diluted with rose oil and applied upon the bregma in the form of a pledget, and constantly changed. (Temkin 1956: 124)

Pliny cited siriasis in several places with a primarily magical treatment:

(Book 22, ch. 29) An application of the leaves [heliotropium] cures the infantile catarrhs that are called siriasis ["dog star fever"], and also convulsions, even though caused by epilepsy (Pliny 1949–1962: vol. 6, 335). (Book 30, ch. 47 [The Troubles of Babies]) The inflammation of babies called siriasis is cured by the bones found in dog's dung worn as an amulet (Pliny 1949–1962: vol. 8, 365). (Book 30, ch. 48) For siriasis in babies a very efficacious cure is a frog tied as an amulet back to front on the infant's skull moistened with a cold sponge. The sponge is said to be found dry afterwards (Pliny 1949–1962: vol. 8, 549).

Kay (1986, pers. com.) wrote:

Esteyneffer came to Mexico in about 1692 and spent some time in the Jesuit college there, where [he] would have had the work of Sahagún, Hernández, and others available to him. Further, he had a Tarahumara curandero informant from whom he learned some indigenous remedies. However, his principal sources were the ancient writers of the Hippocratean-Galenic tradition. He followed the Galenic mode, as did the 16th and 17th Century writers whom he studied. These include Sahagún and Hernández, both of whom were educated following Pliny's *Historia Naturae*.

Several points should be made with respect to Kay's statement. Esteyneffer could not read Nahuatl and so could have used only the Spanish version of Sahagún's *Historia Natural*, which was sent to Spain in 1580, copied several times after 1793, and first published in 1820 (d'Olwer and Cline 1793: 194–200). One cannot assume a priori that Esteyneffer, a Jesuit, would have had access in the New World to a Spanish version of a work by a Franciscan. We know that a copy of Hernández's work was left in Mexico and might have been available to Esteyneffer. However, Hernández did not mention caída de mollera at all and made only one reference to siriasis: "[Tomato is used] for the irritation of children called siriasis mixed with rose oil [an olive oil extract of rose petals]" (Hernández 1959: vol. 1, 228). Esteyneffer did not mention either siriasis or tomatoes. Further, as described below, Esteyneffer's treatments for caída de mollera did not resemble Pliny's treatments of siriasis so that Pliny's influence is not evident.

Kay (1986, pers. com.) cited additional references to siriasis ranging from Pliny to sources published after Esteyneffer's work. The crucial point, however, is whether or not Esteyneffer had access to these works or was influenced by them. It is not sufficient to show that a source described particular view somewhere, some time. Since Esteyneffer's *Florilegio Medicinal* was claimed to be *the* source of Mexican and Southwestern folk medicine, the burden is upon the claimants to demonstrate that the sources cited by Kay as influential were available to and used by Esteyneffer to prepare his work. As pointed out above, Esteyneffer's editor listed none of these sources.

Even if siriasis were generally known in sixteenth and seventeenth century Spain, this does not make it the precursor of caída de mollera as it is known today. Siriasis supposedly was an infection that led to such a high fever that dehydration caused the fontanelle to sink. Caída, however, was more likely associated with dehydration due to diarrhea. The widespread incidence of caída in Latin America and its high frequency in Trotter's survey for some locations in the Southwest support such an association much more than one with a brain inflammation, which although serious is less common. Epidemiologically, diarrhea is the most serious problem of poor Latin American children, particularly when combined synergistically with malnutrition (Bryant 1969: 38, 39; Carpenter 1982). Kay (1986, pers. com.), speaking of siriasis, said that "The distinctive features of the disease were not only the visible depression, but primarily the heat and dryness that caused it, which incidentally is the way dehydration is viewed in biomedicine today." However, Trotter (Trotter, Garza, and Garza n.d.) conducted an experiment that disputed this conclusion. Analyzing the symptoms reported in his survey, he found that they fell into three clusters, groups of ten or more informants presenting common symptoms. The symptom clusters were submitted for blind diagnosis to physicians familiar with the region. Two of the three clusters were diagnosed as severe gastroenteritis leading to dehydration. The third was tentatively identified as an upper respiratory infection, with the proviso that meningitis must first be ruled out. The description of acute infectious gastroenteritis in the *Merck Manual* (Berkow 1977: 1077–1079) fits the folk symptomatology of caída de mollera very well:

> [symptoms] . . . Lethargy, anorexia, fever, oliguria, and substantiated weight loss are historical indications of dehydration. . . . some signs may not be present until dehydration becomes critical. These include warm, dry skin with poor tissue turgor, a sunken anterior fontanel, sunken eyes with absent tearing . . . weak or absent sucking and lethargy.

The classical descriptions of siriasis, cited by Kay, do not mention a connection with infantile diarrhea, but infectious gastroenteritis has been a problem ever since humans settled into permanent communities. It seems logical that the connection between this infection and sunken fontanelle would be much more obvious than the occurrence of a fever severe enough to dehydrate a child in the absence of diarrhea.

An important component of Foster's argument for a Spanish origin of humoral theory was its wide distribution across Latin America, although we have seen in the case of susto that relative ubiquity does not prove such origin. Nevertheless, if caída had a Spanish origin one would expect to find it everywhere in Latin America, particularly in Peru and Chile where Foster (1953: 215) claimed that Spanish folklore and Spanish folk medicine persisted more than in any other Latin American country. In fact it is completely absent in Peru and Chile. The standard source on Peruvian folk medicine (Valdizán and Maldonado 1922) has extensive descriptions of susto and *ojeo* ("evil eye") and their cures but no mention of caída de mollera. More recent reports on Peru (Hubi 1954; Bastien 1983, pers. com.; Manheim 1984, pers. com.), including an extensive study of beliefs on etiology and treatment of infantile diarrhea in Lima (Escobar, Salazar, and Chung 1983), did not mention caída de mollera. Also, an article on Spanish survivals in folk medicine in Chile (Alvarez et al. 1983) did not include caída de mollera.

If Esteyneffer were the source of beliefs concerning caída de mollera, one would expect that treatments today would reflect his prescription for fallen fontanelle:

If the fontanelle of the child is fallen, the mother should put breast milk in the fontanelle itself and she will see it visibly rise. Or put the child's head into a vessel of lukewarm water to the depth of the nose, but don't let the water get into the nose, and lift out suddenly, repeating it several times by which the water will suck the fontanelle out. After this treatment, put a plaster on the fontanelle made out of incense powder or from *copal* made into a paste with a good bit of beaten egg white placed on a cloth. It should be applied lukewarm. (Esteyneffer 1978: vol. 1, 441)

The data in table 8.8 shows that Esteyneffer's treatments are not the preferred modalities today. Mother's milk, incense, and copal poultices are not cited at all. Dipping the head into water was reported by 1.3 percent and 2.9 percent of Trotter's three-state and Rio Grande samples, respectively, while an egg poultice was reported by 3.6 percent and 5.1 percent, respectively. The overwhelmingly predominant remedies all over Mesoamerica—pushing up on the palate and turning over and shaking the child—were not mentioned by Esteyneffer. The case for siriasis as the precursor also is not supported. Although Esteyneffer did not describe an etiology, the remedies he prescribed (hydraulic suction) are consistent only with a mechanical model. If siriasis and caída were one and the same, Hippocratic-Galenic theory would require cooling remedies to counteract the heat of the fever. This is not the case. The modern etiology for caída de mollera also is entirely mechanical. It is possible to lose an etiological explanation and retain a remedy, as shown for Tlaloc's diseases, but modern folk medicine has retained neither the name, the etiology, nor the remedies for siriasis.

The claims of a purely European origin for caída de mollera is thus not tenable. However, another origin satisfies all the requirements for proof defined above. Assuming the Aztec concept of tonalli (discussed in chap. 2) as a contributing source can explain the geographical distribution of belief in the caída syndrome and the predominant cures. According to the Aztecs, caída de mollera was an extreme case of loss of tonalli in children. Children were particularly susceptible because their fontanelles were not yet fully closed (López Austin 1988: vol. 1, 206) and were located at a sensitive site for loss of tonalli. One of the earliest documents in Nahuatl collected by Sahagún, the *Primeros Memoriales*, described the functions of the *teapahtiani* ("healer of the fontanelle"):

The TEAPAHTIANI thus cures little children: she hangs him upside down, she shakes his head from one side to the other and
she pushes on his palate. Some of them attract [the spirit] with
their breath, and also push the child's palate with cotton which
they stuff in. Some get well with this, others do not. This resembles the method in which they puncture [the palates] of little
children, from which they soon die—or they rub them with salt
or they press repeatedly [*papachoa*] the little children with
tomato (Garibay 1943–1946: 242).

The treatments prescribed in this passage, in contrast to those in
Esteyneffer, are still the treatments of choice today. As in the syncretism observed in Tlaloc's diseases, the pre-Columbian explanation of loss of tonalli has been replaced by a mechanical etiology,
but the Aztec remedy is still being applied to an Aztec ailment.

Early Mesoamerica was a cultural unit sharing a common
worldview, religion, and ideas of human physiology and health
(Ortiz de Montellano 1989b). If the real source of caída de mollera
was a loss of tonalli in children, then the cultural unity of Mesoamerica explains the omnipresence of the illness in Mexico,
Guatemala, Honduras, and part of El Salvador, which now occupy
the area. Countries like Chile and Peru that did not share Aztec
beliefs and found no forerunner of caída in Spanish medicine have
no reason today to have caída as part of their folk medicine.

The cure for the loss of tonalli, quoted above, was written in
Nahuatl between 1562 and 1575 (about forty years after the Conquest). As we have discussed, Foster claimed that such sources
could nevertheless have been contaminated by Spanish influence.
In previous chapters we argued that tonalli was so integral to the
Mesoamerican worldview that it is inconceivable that the Spanish
introduced it into Sahagún's work. In this case, however, an additional proof of pre-Columbian authenticity can be adduced—
correspondence between early written descriptions in sources
such as Sahagún and undisputed pre-Columbian hieroglyphs. As
we have stated, such correspondence is one of the most reliable
indications of the absence of European contamination. The standard Mesoamerican pictograph for the capture of an enemy
shows the captor grasping the hair on the crown of the captive's
head, that is, taking possession of his "vital force," or tonalli (fig.
8.2). This pictograph can be seen in Aztec manuscripts such as the
Mendoza Codex (Cooper Clark 1938: fol. 1, 64, 68) and the *Codex
Boturini* (1964: 18), as well as in manuscripts of other cultures,

Figure 8.2.   Symbol for taking a captive, who is grasped by the hair on the crown of his head. (Cooper Clark 1938. *Codex Mendoza*, plate 1)

such as the Mixtec—*Codex Nuttall* (Miller 1975: fol. 3, 4, 21, 76). Its pre-Columbian origin is further verified by its appearance on stone monuments that clearly predate the Spanish Conquest, for example, the Aztec *Tizoc* Stone, in the Mexican National Museum of Anthropology, and the Maya Stela #11 from Yaxchilan, in the Peabody Museum at Harvard. The use of this glyph by the Maya in Guatemala and other areas shows that similar concepts of tonalli existed throughout Mesoamerica and that caída de mollera is grounded on a widely diffused pre-Columbian Mesoamerican concept.

Thus, we have again found an Aztec disease being treated with Aztec remedies, as we did in the modern use of yauhtli and iztauhyatl, but whose etiological explanation was lost during the Colonial period. This pattern of syncretism may well be a general phenomenon, but research is needed to discover further examples. In the case of caída de mollera, the etiology involving loss of tonalli has been replaced by mechanical analogues, such as fall of the palate, whose origin is uncertain. They were not a part of orthodox European medical theory based on humoral principles. Present beliefs about caída de mollera may be due to syncretism, but their predominant source was the native Aztec tradition. Further research may uncover the missing links in the transmission of this syndrome and other examples of reverse acculturation in Mesoamerican folk medicine.

The conclusion that can be derived from this chapter is that much of modern Mexican folk medicine derives from a syncretism of European concepts with preexisting native concepts. Broad generalizations are not warranted. The proportions of each cultural strand vary from case to case, which should be examined individually. No one text, such as Esteyneffer, can be nominated as the fountainhead of Latin American folk medicine.

Nutritional Values and Amino Acid Composition of Aztec Foods

Nutritional Values of Aztec Foods per 100 Grams

Food[a]	CAL	PROTEIN g	CA mg	P mg	VIT. A mg	THIAMINE mg	RIBOFLAVIN mg	NIACIN mg	VIT. C mg	FE mg	REF[b]
Armadillo	181	29.0	30	208	0	0.1	0.4	6.01	0	10.9	1
Ahuauhtli (dry)		63.8	104	693	—	0.41	0.91	11.4	—	9.5	3, 4
Axayacatl (dry)		53.8	613	759	—	0.54	2.02	11.49	—	48	4
Charales (dry)		61.9	4160	2640	—	0.40	0.10	6.0	—	18.1	4, 5
Chicatana ant		58.3									2
Deer	155	29.5	20	264	0	0.37	0.28	7.43	—	3.5	1
Duck	115	23.3	7	203	—	0.15	0.39	11.7	—	2.8	1
Escamoles (dry)		66.9									2
Grasshopper (dry)		62.0									2
Iguana	113	24.4	25	252	0.23	0.05	0.24	8.18	—	3.4	1
Jumiles		32.3	78	285	—	0.32	1.28	3.76	—	10.1	5
Maguey worm (white, dry)		16.6	142	140	—	0.42	0.58	3.0	—	4.3	5
Maguey worm (red, dry)		71.0									2
Rabbit	99	18.4	25	178	—	0.06	0.14	9.18	—	1.2	1
Turkey	268	20.1	23	320	tr	0.09	0.14	8.0	—	3.8	1

(continued)

Nutritional Values of Aztec Foods per 100 Grams (Continued)

Food[a]	CAL	PROTEIN g	CA mg	P mg	VIT. A mg	THIAMINE mg	RIBOFLAVIN mg	NIACIN mg	VIT. C mg	FE mg	REF[b]
Avocado	120	1.6	11	40	0.06	0.07	0.12	1.6	13	0.7	1
Black bean	343	22.7	134	415	0.01	0.47	0.15	2.09	1	2.3	1
Cactus leaves		2.0	81	20	0.25	0.02	0.08	0.24	12	2.34	4
Cactus fruit		1.31	90	31	—	0.02	0.02	0.29	31	1.1	4
Chia	463	15.6	518	518	0.01	0.38	0.13	3.74	—	10.0	1
Epazote	43	5.0	342	59	1.8	0.17	0.30	0.93	99	8.6	1
Chile	81	3.9	33	67	0.75	0.08	0.39	1.44	89	1.8	1
Chile	84	14.6	200	180	4.50	0.29	1.06	13.8	22	14.3	1, 4
Corn	358	8.4	11	121	0.15	0.48	0.10	1.9	—	1.5	1
Tortilla	206	5.6	158	122	0.07	0.12	0.05	1.02	—	2.5	1
Corn		8.3	8	235	0.15	0.34	0.08	1.64	—	2.3	4
Tortilla		5.8	111	184	0.06	0.19	0.06	0.96	—	2.2	4
Pozole		6.0	59	158	0.28	0.16	0.12	0.76	—	2.3	4
Hediondilla		5.1	150	46	2.72	0.09	0.36	1.10	89	4.3	6
Huauhtli	391	15.3	490	455	—	0.14	0.32	1.0	5	17.4	7
Huauhtli lvs.	36	3.5	267	67	3.7	0.08	0.16	1.4	80	3.9	8

Lengua de vaca	55	1.6	93	29	1.33	0.07	0.19	0.61	32	5.2	6
Mallow	294	6.0	414	120	4.6	0.27	0.56	1.72	120	23.9	1
Mezquite		11.0	140	57	0.15	0.05	0.05	0.95	—	0.1	1, 10
Garambullo			92	27	0.38	0.03	0.01	0.47	40	4.4	2
Purslane	25	0.3	91	34	0.63	0.02	0.11	0.55	25	0.8	1
Pulque	20.4	0.44	10	10	—	0.02	0.02	0.03	6	0.7	9
Squash	20	1.2	28	29	0.28	0.05	0.09	1.0	25	0.4	8
Tecuitlatl (dry)	600	70	131	894	—	5.5	4.0	11.8	10	58	14, 11
Tomato	22	1.1	13	27	0.27	0.06	0.04	0.7	23	0.5	8
Yucca	28	0.9	340	23	0.01	0.01	0.04	0.3	25	0.9	1

See notes after following table.

Amino Acid Composition of Foods mg/g Nitrogen

FOOD[a]	ILE	LEU	LYS	PHA	MET	THR	TRY	VAL	SCORE[c]	REF[b]
FAO pattern	270	106	270	180	144	180	90	270	100	13
Egg	428	565	396	368	196	310	106	460	100	13
Cow's milk	407	630	496	311	154	292	90	440	78	13
Beef	301	507	556	275	169	287	70	313	78	12
Poultry	334	460	497	250	157	248	64	318	71	12
Pork	356	563	625	288	188	319	85	388	94	12
Iguana	553	607	590	705	163	468	67	334	74	12
Rabbit	340	410	510	220	190	320	95	400	100	12
Jumiles	264	495	199	231	109	270	6.4	469	7	2
Maguey worm red	328	508	315	257	51	302	39	392	35	2
Maguey worm white	315	334	231	238	64	212	58	302	44	2
Axayacatl	379	514	276	206	103	283	103	354	72	2
Ahuauhtli	321	514	225	219	96	257	71	386	67	2
Chicatana ant	341	514	315	264	122	276	39	411	43	2
Escamoles	315	489	273	251	116	270	51	386	35	2
Grasshopper	296	411	334	231	51	315	64	347	35	2
Amaranth	212	334	321	244	129	206	51	244	57	7

Amaranth	444	514	527	302	161	276	58	386	64	7
Corn	293	827	179	284	117	249	38	327	42	13
Navy bean	358	541	460	347	64	274	58	379	44	13
Tecuitlatl	386	514	296	321	141	296	90	418	98	14
Tecuitlatl	388	516	295	320	88	293	90	417	61	15
Mezquite	189	106+	230	180+	58	126	50	189	40	17
Pozole	332	644	255	276	95	366	46	291	51	16
Charales	341	476	514	199	135	283	58	334	64	5

[a]Identification of species: armadillo (*Dasypus sexcinatus*), charales (*Corisella* sp.), ahuauhtli (*Corisella* sp. eggs), axayacatl (*Corisella* sp.), charales (*Chirostoma* sp.), chicatana ant (*Atta mexicana*), deer (*Cervus elaphus*), duck (*Anas domesticus*), escamoles (*Liometopum apiculatum*), grasshopper (*Sphenarium histro*), iguana (*Lacerta iguana*), jumiles (*Atizies* sp.), white maguey worm (*Aegiale hesperiaris*), red maguey worm (*Cossus redtenbachi*), rabbit (*Osyetolofagus cuniculus*), turkey (*Meleagrio gallopavo*), avocado (*Persea americana*), black bean (*Phaseolus vulgaris*), cactus (*Opuntia* sp.), chia (*Salvia hispanica*), epazote (*Chenopodium ambrosoides*), chile (*Capsicum fructescens*), corn (*Zea mays*), tortilla (*Zea mays*), pozole (*Zea mays*), hediondilla (*Chenopodium album*), huauhtli (*Amaranthus hypochondriacus*), lengua de vaca (*Rumex crispus*), mallow (*Malva parviflora*), mezquite (*Prosopis juliflora*), garambullo (*Myrtilocactus geometricans*), purslane (*Portulaca oleracea*), pulque (*Agave* sp.), squash (*Cucurbita pepo*), tecuitlatl (*Spirulina maximus*), tomato (*Lycopersicon esculentum*), yucca (*Yucca elephantipes*).

[b]1. Flores et al. 1960; 2. Ramos de Elurdoy 1982; 3. Cravioto 1951; 4. Cravioto, Massieu, and Guzmán 1951; 5. Massieu et al. 1951; 6. Cravioto et al. 1945; 7. Cole 1979; 8. Consumer and Food Economics Research Division 1963; 9. Gonzalvez de Lima 1956; 10. Gupta, Gandhi, and Tandon 1974; 11. Santillán 1982; 12. Bresani 1976; 13. Chaney and Ross 1971; 14. Nakamura 1982; 15. Gordon 1970; 16. Cravioto et al. 1955; 17. Felker and Bandurski 1979.

[c]Chemical score is calculated by dividing the amount of limiting amino acid by the amount found in the Food And Agriculture Organization (FAO) reference protein.

Empirical Evaluation of Aztec Medicinal Herbs

Aztec Medicinal Herbs[a]

BOTANICAL NAME	AZTEC NAME	NATIVE USES	RELEVANT CHEMICALS	EFFECTIVENESS	EMIC	ETIC
Achillea millefolium (Compositae)	tlalquequetzal	coughs, sores, scabies, body aches	tannins, cineole, chamazulene	coughs / sores / aches	+ / + / +	3 / 3 / 3
Adiantum poiretti (Polypodiaceae)	tequequetzal	wounds, fever	tannins, mucilage	fever / wounds	+ / +	1 / 2
Agave atrovirens (Agavaceae)	metl	wounds	polysaccharides, saponins	wounds	+	4
Annona cherimolia (Annonaceae)	quauhtzapotl	diarrhea	tannins, annonaine	diarrhea	+	3
Arctostaphylos tomentosa (Ericaceae)	tepetomatl	diuretic, inflammation	tannins, arbutin, quercetin	diuretic / inflammation	+ / +	3 / 2
Argemone mexicana (Papaveraceae)	chillazotl	purgative, fever, inflamed eyes	berberine, protopine, sanguinarine	fever / eyes	+ / +	1 / 2
Artemisia mexicana (Compositae)	iztauhyatl	fever, colic, phlegm	camphor tannins, thujone	colic	+	2
Asclepias linaria (Asclepiadaceae)	tlalacxoyatl tezonpatli	ulcers, mange, laxative	asclepiadin, asclepion	skin	+	1
Bixa orellana (Bixaceae)	achiotl	fever, diuretic, diarrhea	tannins, bixin, tomentosic acid	fever / diarrhea / diuretic	+ / + / +	1 / 2 / 1

(continued)

Aztec Medicinal Herbs[a] (*Continued*)

BOTANICAL NAME	AZTEC NAME	NATIVE USES	RELEVANT CHEMICALS	EFFECTIVENESS	EMIC	ETIC
Bletia campanulata (Orchidaceae)	tzacuxochitl	against fear	mucilage	fear	–	–
Bocconia frutescens (Papaveraceae)	cocoxihuitl	constipation, diuretic, inflammation	protopine, chelerythrine, dehydrosanguinarine	laxative diuretic inflammation	+ + +	2 2 2
Bromelia pinguin (Bromeliaceae)	mexocotl	heat blisters in mouth	pinguinain	blisters	+	2
Buddleia americana (Loganiaceae)	zayolitzcan	fever, diuretic, wounds	aucubin, catalpol	fever diuretic wounds	+ + –	1 2 1
Calliandra anomala (Leguminoseae)	tlacoxilo-xochitl	nosebleed, fever, cough, dysentery	tannis, calliandreine, saponins	fever dysentery cough	– + +	– 2 2
Carica papaya (Caricacaea)	chichihual-xochitl	rash, ulcers, digestive	papain, carpain, papase	rash digestive	+ +	2 3
Casimiroa edulis (Rutaceae)	cochiztzapotl	sedative, soporific	N,N-dimethylhistamine, casimiroin, fagarine	sedative soporific	+ +	3 3
Cassia occidentalis (Leguminosae)	totoncaxihuitl	purgative, fever	chrysophanic acid, rhein	purgative	+	3
Cassia alata (Leguminosae)		astringent, inflammation	tannis, chrysarobin	astringent inflammation	+ +	3 3

Plant (Family)	Nahuatl name	Uses	Constituents			
Chiranthodendron pentadactylon (Sterculiaceae)	macpalxochitl	inflammation, pain in groin	quercetin, sitosterol, cyanidin-3-glucoside	inflammation	+	2
Commelina pallida (Commelinaceae)	matlaliztic	hemostatic	tannins, gallic acid	hemostatic	+	2
Croton sanguifluum (Euphorbiaceae)	ezquahuitl	diarrhea, astringent	cathechins, cathecols, saponins, tannins	diarrhea	+	2
Datura stromonium (Solanaceae)	tlapatl	fevers, locally for chest pain and gout, hallucinogen	atropine, scopolamine, hyoscyamine	fever hallucinogen chest pain	? + +	– 4 3
Eryngium aquaticum (Umbelliferae)	ocopiaztli	fever, diuretic, inflammation	rosmarinic acid, sapogenins, chlorogenic acid	fever diuretic	+ +	1 1
Euphorbia calyculata (Euphorbiaceae)	cuauhtepatli	purgative, skin ailments	salicylic acid, euphorbol	purgative skin	+ +	3 1
Euphorbia pulcherrima (Euphorbiaceae)	cuetlaxochitl	increase milk in nursing woman	germanicol, kaempferol, pulcherrol	milk	magic	–
Eysenhardtia polystachia (Leguminosae)	coatli	diuretic, fever, hiccoughs	dehydrorotenone, β-sitosterol	fever diuretic	+ +	– 1
Gossipium herbaceum (Malvaceae)	ichcaxihuitl	skin sores, animal, snake bites	gossipol, quercetin,	skin bites	– –	– –

(continued)

Aztec Medicinal Herbs[a] (Continued)

BOTANICAL NAME	AZTEC NAME	NATIVE USES	RELEVANT CHEMICALS	EFFECTIVENESS	EMIC	ETIC
Guaiacum sanctum (Zygophyllaceae)	matlatlqua-huitl	fever, infections, syphylis, kidney illness, sores	guaiacol, resin, nor-dihydroguajaretic acid	fever infections kidney	+ + +	— 2 2
Helianthus annuus (Compositae)	chimalacatl	fever	saponins, quercimeritrin	fever diuretic	+ +	— 2
Ipomoea purga (Convolvulaceae)	cacamotic	purgative, heat in heart	convolvuline	purgative heat in heart	+ +	3 —
Iresine calea (Amarantaaceae)	tlatlanquaye	diarrhea, epilepsy	tlatlancuayin iresin	diarrhea epilepsy	— —	— —
Jathropa curcas (Euphorbiaceae)	ayohuactli axquauitl	emetic, purgative	sapogenin, tannins, resin	purgative	+	2
Liquidambar styraciflua (Hamamelidaceae)	ocotzotl	rash, toothache, stomach tonic	storenin, styrol, cinnamic acid esters	rash toothache	+ +	2 2
Lobelia laxiflora (Campanulaceae)	chilpanton	nosebleed, cough, emetic	lobeline, lobelanine	cough emetic	+ +	3 3
Mirabilis jalapa (Nyctaginaceae)	atzomiatl tlaquilin	diarrhea, in-fected ears	trigonellin	diarrhea earache	+ +	1 1
Montanoa tomentosa (Compositae)	cihuapatli	oxytocic, diure-tic	zoapatanol, montanol, kaurenedienoic acid	oxytocic	+	4

Species (Family)	Name	Uses	Constituents	Effect		No.
Passiflora jorullensis (Passifloraceae)	coanenepilli	diuretic, fever, diaphoretic, snake bites, pain	harmine, harman, passicol	fever snake bite	+ magic	— —
Perezia adnata (Compositae)	pipitzahuac	purgative, fever, emetic	gallic acid, perezone, β-pipitzol	fever purgative emetic	+ + +	— 3 3
Persea americana (Lauraceae)	auacatl	astringent, sores	serotonin, tryptamine unsaturated heptadecatriols	sores	+	2
Piper amalago (Piperaceae)	mecaxochitl	fever; diuretic, diarrhea	safrole, aporphine alkaloids	fever diuretic	+ +	1 1
Pithecellobium dulce (Leguminosae)	quamochitl	astringent, ulcers, sores	quercetin, kaempferol, tannins, pithecolombine	astringent sores	+ +	3 1
Plantago mexicana (Plantaginaceae)	acaxilotic	fever, emetic, laxative	acubin, mucilage	fever laxative	+ +	— 3
Plumbago pulchella (Plumbaginaceae)	tlepatli, itzcuinpatli	diuretic, colic, gangrene	plumbagin	skin	+	2
Plumeria rubra (Apocynaceae)	cacaloxochitl	purgative, fear	kaempferol, quercetin, plumieride	fear purgative	+ +	— 3
Polianthes tuberosa (Agavaceae)	omixochitl	fever, rash, diarrhea	tigogenin, hecogenin, benzoic acid	fever diarrhea	— —	— —
Polygonum hidropiper (Polygonaceae)	achilli, achilto	hurt feet, headache	gallic acid, quercetin; hyperin	hurt feet headache	+ —	1 —

(continued)

Aztec Medicinal Herbs[a] *(Continued)*

BOTANICAL NAME	AZTEC NAME	NATIVE USES	RELEVANT CHEMICALS	EFFECTIVENESS	EMIC	ETIC
Prosopis juliflora (Leguminosae)	mizquitl	eye medicine, excessive menses	serotonin, tryptamine, prosopine, gum	eyes	+	1
Prunus capulli (Rosaceae)	capolin	diarrhea, inflammation	tannins, gallic acid, pectin	diarrhea inflammation	+ +	3 2
Psidium guajava (Myrtaceae)	xalxocotl	dysentery, mange	tannins, guijaverin, caryophellene	dysentery skin	+ +	3 2
Quercus spp. (Fagaceae)	auaquauitl	dysentery, tiredness of public officials	tannins	dysentery tiredness	+ magic	3 –
Rhamnus serrata (Rhamnaceae)	tlalcapulin	dysentery	chrysophanic acid	dysentery	–	–
Rumex mexicana (Polygonaceae)	amamaxtla	purgative, diuretic, astringent	tannins, chrysophanic acid, gallic acid	purgative diuretic	+ +	3 2
Salix lasiopelis (Salicaceae)	quetzalhuexotl	fever, dysentery	salicin, picein	fever dysentery	+ +	3 1
Sambucus mexicana (Caprifoliaceae)	xumetl	purgative, diuretic	kaempferol, tannins, astragalin, sambunigrin	laxative diuretic	+ +	3 3
Schoenocaulon coulterii (Liliaceae)	zozoyatic	headache, kill flies and mice	cevadine, jervine, veratridine	headache kills mice, flies	+ +	– 2
Smilax aristolochiaefolia (Smilacaceae)	mecapatli	diaphoretic, diuretic; pain in joints	sarasapogenin, parillin, sitosterol	diaphoretic diuretic	+ +	2 2

Species (family)	Nahuatl name	Uses	Chemicals	Category		
Stevia salicifolia (Compositae)	tonalxihuitl	fever, pimples, local pain	saponins, bisabolene	fevers	+	1
Tagetes erecta (Compositae)	cempohual-xochitl	diaphoretic, purgative, dropsy	quercetagitin, patuletin, tagetone	diaphoretic	+	2
Tagetes lucida (Compositae)	yiauhtli	fever, fear, di-uretic, de-mentia, lightning strike	tagetone, quercetagritin, kaempferol, tagetiin	fever diuretic fear, lightning	+ + +	— 3 —
Talauma mexicana (Magnoliaceae)	yolloxochitl	comforts heart, fever, diuretic, sterility	talaumine, aztequine, costunolide	heart fever diuretic	+ + +	2 — 2
Teloxys ambrosioides (Chenopodium ambrosioides) (Chenopodiaceae)	epazotl	dysentery, an-thelmintic	ascaridole, p-cymene, camphor, 1-limonene	anthelmintic	+	3
Thalictrum hernandezii (Ranunculaceae)	costic patli	eye medicine, diuretic	berberine, magnoflorine	eyes	+	2
Theobroma cacao (Sterculiaceae)	cacahua-quauitl	with rubber for diarrhea, ex-cess makes dizzy	theobromine, tannins	dizziness	+	2
Turbina corymbosa (Rivea corymbosa) (Convolvulaceae)	ololiuhqui	hallucinogenic, locally for gout	ergonovine, ergine	hallucination gout	+ ?	4 1

a) References for chemicals and identification can be found in Ortiz de Montellano (1974/75: 1975).

NOTES TO APPENDIX B

Some types of compounds occur several times and will be discussed together, before individual species are covered. Tannins are complex mixtures of polyphenols, which do not crystallize, widely distributed in the plant kingdom. Some of the simpler hydrolysis products are gallic and ellagic acids and cathecol. Tannins precipitate proteins from solution causing an astringent action in living tissues. This astringency is useful to halt bleeding and, in the gastrointestinal tract, to halt diarrhea. The polyphenolic nature of tannins makes them antiseptic when applied locally and

> In the treatment of burns, the proteins of the exposed tissues are precipitated forming a mildly antiseptic, protective coat under which the regeneration of new tissue may take place (Claus, Tyler, and Brady 1970: 131).

Saponins and sapogenins are generally irritating substances whose effects depend on the site of irritation. For example, they can promote expectoration in the treatment of coughs and have a diuretic effect on the kidneys. In general, saponins also have a strong diuretic action as well as some diaphoretic and laxative properties (Tyler 1987: 206). Anthraquinone glycosides yield anthraquinones such as chrysophanic acid, chrysarobin, and rhein, and are potent irritant purgatives (Claus, Tyler, and Brady 1970: 96–108). Glycosides of flavones, such as quercetin and patuletin, are diuretic, while kaempferol has a laxative effect (Steinegger and Haensel 1963: 169, 174).

Achillea millefolium. Camphor is used as a liniment for its local anesthetic and antipruritic effects and functions as an expectorant. Chamazulene is an effective anti-inflammatory. The plant has been used by different native American tribes for coughs, skin eruptions, swelling, and bruises (Moerman 1986: 9–11). Thompson (1978: 184) listed its effects as "antiseptic, antispasmodic, expectorant, astringent and anti-flammatory."

Adiantum poiretti (Sahagún 1950–1969: book 11, 196; Hernández 1942–1946: vol. 2, 340). Tannins are hemostatic. The plant is used as a diuretic in Guatemala (Orellana 1987: 174) and by the Iroquois (Moerman 1986: 11, 19). Diuretic action satisfies Aztec criteria for a fever remedy. *A. lucida* is used in the West Indies as a febrifuge (Ayensu 1981: 152).

Agave atrovirens. The effectiveness of *Agave* sap for wounds has been discussed.

Annona cherimolia. Tannins are effective against diarrhea due to their astringent action. *A. reticulata* is used in the West Indies as an astringent and against diarrhea (Ayensu 1981: 40). Aqueous extracts of *A. cherimolia* affected the intestinal and uterine tissues of guinea pig in vitro. Uterine contractility was due to a steroid glycoside (Lozoya 1978). This effect was not listed by the Aztecs although the plant is used as a folk abortifacient in Mexico (Quezada 1975a), and other Annonas have been used elsewhere for the same purpose (Farnsworth et al. 1975).

Arctostaphylos tomentosa. Arbutin hydrolyses to hydroquinone in alkaline urine where it acts as a mild astringent and antiseptic. Ursolic acid and isoquercetin, which have a diuretic action at dilutions of 1:100,000, have been isolated from the plant (Spoerke 1980: 30). *A. uva-ursi* was used by the Cherokee as a diuretic (Moerman 1986: 53–54) and was used formerly in the United States as a diuretic, antiseptic, and astringent, providing topical relief to swollen areas (Morton 1977: 368).

Argemone mexicana. The oil and seeds of this plant are purgative and emetic, as is protopine. This validates its use against fevers by the Aztec. *A. mexicana* has been used in Jamaica and Curazao against fevers and in Yucatan and Venezuela for eye inflammation (Morton 1981: 241–242). Berberine is a mild local anesthetic of mucous membranes and has been used in over-the-counter medications for irritated eyes (Orellana 1987: 178).

Artemisia mexicana. Thujone is a component of oil of wormwood, formerly used medicinally as an anthelmintic and colic reliever. Camphor is a mild irritant, stimulant, and colic reliever. The plant has been used in Guatemala for colic (Morton 1981: 909).

Asclepias linaria (Sahagún 1950–1969: book 10, 144; book 11, 141, 156; de la Cruz 1964: 245; Hernández 1942–1946: vol. 1, 42). Various *Asclepias* species have been used by a number of Native American tribes as an antiseptic to treat sores, ringworm, and warts and as laxatives (Moerman 1986: 75–79). Morton (1981: 687–690) reported that *A. curassavica* was used in Yucatan and Guatemala for skin diseases and as a purgative. *A. curassavica* contains asclepiadin and was found experimentally to cause vomiting and diarrhea (Lozoya et al. 1978).

Bixa orellana. This is listed as an astringent, fever reducer, and antidysenteric in the Mexican pharmacopoeia. The shell has a vermifugic and anthelmintic action. The plant is used in Guatemala for fevers and as a diuretic (Orellana 1987: 181). It is also used as a diuretic in Venezuela and Trinidad and for diarrhea in Yucatan and Venezuela (Morton 1981: 572–573). Its diuretic effect validates it emically as a febrifuge.

Bletia campanulata (Sahagún 1950–1969: book 10, 87; book 11, 211; de la Cruz 1964: 283; Hernández 1942–1946: vol. 2, 775). Mucilage would be a bulk laxative (Claus, Tyler, and Brady 1970: 67–78) and thus valid according to the Aztecs, since fear causes the accumulation of phlegm, which would have to be expelled.

Boccania frutescens. Chelerythrine and sanguinarine are active local irritants and would be expected to be laxative and diuretic. Aqueous extracts have diuretic, antimicrobial, and anti-inflammatory activity. The sap is purgative, and the plant is used for ulcers and skin eruptions in Haiti and Costa Rica (Morton 1981: 242–244). It was used for mange and against skin parasites in Colombia (von Reis and Lipp 1982: 88).

Bromelia pinguin. Pinguinain is a proteolytic enzyme and also reduces swellings. The juice is used in the West Indies for thrush and other ulcerations of the mouth (Ayensu 1981: 62).

Buddleia americana (Sahagún 1950–1969: book 10, 35; book 11, 110, 168; de la Cruz 1964: 283; Hernández 1942–1946: vol. 3, 749). The plant contains aucubin and catalpol (Gibbs 1974: vol. 1, 625; Hegnauer 1962–1977: vol. 3, 308–310). Catalpol and aucubin are diuretic and laxative (Sticher 1977) and thus emically would be effective fever medicines. Morton (1981: 664) reported that this species showed marked diuretic action, was used in Guatemala as a diuretic, and was also applied to wounds.

Calliandria anomala. The styptic astringent effect of tannins is helpful in both nosebleed and diarrhea. Saponins function as expectorants in relieving coughs, and their diuretic effect would be considered a treatment of fever by the Aztecs.

Carica papaya. Papain is a digestant that can be used both externally and internally. It is an ingredient in modern digestive preparations (Spoerke 1980: 130; Morton 1981: 596–597). The latex is used against various skin diseases in several countries and against boils and ringworm in the West Indies (Morton 1981: 596–597; Ayensu 1981: 75).

Casimiroa edulis. The Nahuatl name of the plant, "sleeping sweet fruit," refers to its properties. It is widely used in Mexico as a sedative, soporific, and hypotensive. N,N-dimethylhistamine is effective orally and produces a general vasodilation, which lowers blood pressure. This sedative action in larger doses can lead to somnolence or even loss of consciousness (Morton 1981: 372; Lozoya and Lozoya 1982: 130–147). Lozoya found that the presence of dextrose assisted in the absorption of the active compound.

Cassia occidentalis. Chrysophanic acid and its glycosides are purgative, and chrysarobin is used topically for psoriasis and other skin diseases. It is used for skin diseases in the Philippines, North Borneo, and a number of countries in Africa and for inflammation in Ghana and Gambia (Ayensu 1978: 72–75; von Reis Altschul 1973: 116). *Cassia* species are also used as purgatives in Africa, Brazil and North American tribes (Ayensu 1978: 72–75; von Reis Altschul 1973: 116; Moerman 1986: 103–105).

Chiranthodendron pentadactylon. Bye and Linares (1987) felt that this plant was used medicinally by the Aztecs but also had great ritual importance, since both Hernández and Sahagún cited it several times, providing illustrations but only a very sparse and shallow text, indicates that information may have been censored. For a long time, its principal use, occasionally mixed with *Talauma mexicana,* has been in the treatment of heart conditions. Aqueous extracts have shown antiinflammatory activity (Jiu 1966). Luteolin glycosides and quercetin glycosides are diuretic and have shown a stimulating effect on the heart. Extracts were found to cause tissue contractions of the aorta, uterus, and trachea (rabbit); trachea, intestine, and uterus (guinea pig); aorta and intestine (rat); and increased arterial pressure of intact dogs and rabbits (Lozoya et al. 1978).

Commelina pallida. Its use in wound treatment has been discussed.

Croton sanguifluum. Cathechin is astringent, useful for diarrhea, and produces a local cooling effect.

Datura stramonium. Atropine is absorbed through the skin and relieves the pain of muscular rheumatism and gout. Datura was an official drug in the U.S. pharmacopoeia, used particularly as a plaster for chest pains until thirty years ago (Morton 1981: 794–795; Greve 1971: vol. 2, 806). It has been used in Mexico for inflammation, in Yucatan for pain of rheumatism, and by the Rappahanock for fever and chest pain (von Reis and Lipp 1982: 268; Díaz 1979; Morton 1981: 794–796; Moerman 1986: 149–150). Its peripheral cholinergic effects are antiasthmatic, purgative, expectorant, and mydriatic (Díaz 1979). Its use by the Aztecs as a fever remedy would be due to its purgative effects. The plant is a well-known, although dangerous, hallucinogen since its effective dose is close to its lethal dose.

Eryngium aquaticum. The saponins present are diuretic. The plant is used as a diuretic in Guatemala and by the Choctaw and as a febrifuge in the West Indies and Surinam (Moerman 1986: 176; Morton 1981: 646–648; Ayensu 1981: 185). It would have been regarded as effective for fevers by the Aztec.

Euphorbia calyculata. Various *E*. species are used as purgatives due to a resin that resembles the resin of croton oil (Keys 1976: 103–104). Salicylic acid is an active disinfectant, probably superior to phenol in its antiseptic properties. The acid is also widely used in the treatment of various skin diseases, especially chronic eczemas, in which its actions are primarily keratolytic (Lewis and Lewis 1977: 151). *E. hirta* is used to treat warts and tumors in India, the Ivory Coast, and Indonesia (Ayensu 1978: 125).

Euphorbia pulcherrima. Morton (1981: 442–445) reported that this is still used as a galactagogue in Mexico, Guatemala, and El Salvador. The latex is irritating and purgative but none of the constituent chemicals produce the desired effect. It was most probably used in a magical function because its abundant latex has a milky appearance.

Eysenhardtia polystachia (Sahagún 1950–1969: book 11, 140, 141; de la Cruz 1964: 278; Hernández 1942–1946: vol. 2, 520). This plant has not been studied chemically. The reported compounds are not diuretic (Domínguez, Franco, and Díaz-Vivecos 1978). However, it continues to be used as a diuretic and for kidney ailments in Guatemala, Mexico, and southwest Texas (Mellen 1974; Domínguez and Alcorn 1985; Kimber 1981).

Gossipium herbaceum. Neither the cotton fibers nor the chemicals isolated from the plant would be effective for the conditions cited.

Guaiacum sanctum. Guaiacol has been used in veterinary practice as an intestinal antiseptic and as an expectorant. Nordihydroguajaretic acid is an antiseptic and a specific inhibitor of 5-lipooxygenase, which would prevent synthesis of leukotrienes involved in allergic and immune responses, and make it an effective anti-inflammatory agent (Navar 1987, pers. com.). The resin has a local irritating effect on the gastrointestinal tract and stimulates perspiration (Spoerke 1980: 82–83). Water extracts of the plant caused contractions in bladder and uterine tissue of rabbits and guinea pigs and increased the blood pressure of dogs and rabbits (Lozoya et al. 1978). It is widely used today as a diaphoretic in Mexico and Guatemala (Mellen 1974; Morton 1981: 367–368). Its use for the treatment of syphilis, as noted in Hernández and Ximénez may be a European addition. The origin and presence of syphilis are disputed, as mentioned in chapter 5. In the sixteenth century it was widely believed that syphilis had been imported from the New World. It was also believed that God had placed the remedies for particular diseases in the same geographical area where the disease originated, and therefore remedies for syphilis had to be sought in the New World. The Hippocratic-Galenic humoral theory of the time attributed syphilis to an excess of phlegm. According to this theory, a remedy would require removal of phlegm, possibly by excessive perspiration. This etiological concept led initially to enormous importation into Europe of *Guaiacum* and later of sarsaparilla (*Smilax* sp.), which was also a diaphoretic (Crosby 1972: 127). This trade continued until mercury became the treatment of choice. Even though very dangerous because of frequent poisoning, mercury was the sole treatment for this disease until Ehrlich's discovery of salvarsan in 1911. The use of mercury in treating syphilis is another case in which the emic rationale for the use of the remedy is a side effect of the truly active component. One of the symptoms of mercury poisoning is incessant salivation, which at that time would have been interpreted as the removal of phlegm and as a cure by humoral etiological principles (Ransford 1983: 10, 170).

Helianthus annuus. Saponins are diuretic, and the plant's seeds and seed oil were commonly prescribed as diuretics (Morton 1981: 940–941). *H. annuus* has been used by the Pima for fevers and in Yucatan and Costa Rica for fever and rheumatism (Moerman 1986; 217–218; Morton 1981: 940–941).

Ipomoea purga. Convolvuline is a purgative (Morton 1981: 369) used as such in Mexico and Central America (Morton 1981: 701). If the Aztecs believed that heat in the heart was caused by excess phlegm, as was fever, the appropriate remedy would be removal of phlegm.

Iresine calea. This plant is now used in folk medicine in Mexico as a fever remedy, diuretic, and diaphoretic, but would not be effective as a purgative (Morton 1981: 187).

Jathropa curcas (Sahagún 1950–1969: book 10, 90; Ximénez 1888: 58; Hernández 1942–1946: vol. 1, 172). The seed oil is a drastic purgative, while the raw seeds cause vomiting and purgation (Orellana 1987: 213). It is used as a folk purgative in Bolivia, Ecuador, Cuba, and the West Indies (Ayensu 1981: 98; Morton 1981: 448–449; von Reis Altschul 1973: 161).

Liquidambar styracyflua. The balsam storax has been used as a parasiticide ointment in scabies and other parasitic infections. It has antiseptic properties. It has been used in Guatemala for treating wounds, sore gums, and toothache (von Reis Altschul 1973: 97). The Cherokee used it as a salve for wounds, sores, and ulcers, and the Choctaw use it as a poultice for cuts and bruises (Moerman 1986: 264).

Lobelia laxiflora. The action of lobeline resembles that of nicotine in that it first stimulates and then depresses the ganglia of the autonomic nervous system. It also stimulates certain medullary centers, especially the emetic center (Spoerke 1980: 110–111). It was used in American medicine in the nineteenth century as an emetic, expectorant, and anti-asthmatic (Gilman et al. 1980: 587).

Mirabilis jalapa (Sahagún 1950–1969: book 11, 219; de la Cruz 1964; 248; Hernández 1942–1946: vol. 1, 194). The alkaloid trigonelline has a laxative effect (der Marderosian and Liberti 1988: 158). This would have the opposite effect from that desired by the Aztecs. The plant is used in Brazil to relieve diarrhea (Morton 1981: 194–195).

Montanoa tomentosa. This plant has been discussed.

Passiflora jorullensis. A number of *Passiflora* species have been used in folk medicine as diuretics (Morton 1981: 587–595; von Reis Altschul 1973: 201). Harmaline has a marked central nervous system effect, including a decrease in body temperature (Spoerke 1980: 134), which might have been interpreted as a fever cure. Its use as a remedy for snake bites was probably magical since the vine resembles a snake.

Perezia adnata. Whole root extracts have different effects depending on the dose. At lower doses, a purgative effect is due to increased intestinal peristalsis, while emetic effects are produced at higher doses (Lozoya and Lozoya 1982: 60–79). The active compound was found to be an angelic acid ester of hydroxyperezone. The Coahuilla tribe used this species as a purgative (Moerman 1986: 330).

Persea americana. Unsaturated heptadecatrienes and their acetate esters are antibiotic against gram-negative bacteria. Extracts of the seed and peel were found to be antibiotic (Lozoya and Lozoya 1982: 26). It is used against rashes and sores in Guatemala (Orellana 1987: 225).

Piper amalago (Sahagún 1950–1969: book 10, 77, 78, 153; de la Cruz 1964: 274; Hernández 1942–1946: vol. 3, 751). *P. amalago* is used as a diuretic and diaphoretic in Mexico and the West Indies (Morton 1981: 124–125; Ayensu 1981: 148). As such it would have been considered a febrifuge by the Aztecs.

Pithecellobium dulce. The tannins have a local astringent, antiseptic effect. *P. inguis-cati* is used in Curazao on sores as an astringent (Morton 1981: 343).

Plantago mexicana. Several *Plantago* species are cited in the British pharmacopeia for chronic constipation. The effect is due to the bulk laxative effect of the mucilage and to aucubin (Trease and Evans 1978: 354). This effect would fit Aztec etiology as a febrifuge. The plant was used against fever in Puerto Rico and Cuba and by the Costanoan and Rappahanock tribes against fever and constipation (von Reis and Lipp 1982: 282; Moerman 1986: 350–352).

Plumbago pulchella. Plumbagin is active against bacteria, particularly staphylococcus, and has caustic effects externally. "Gangrene," cited in the table, may not be the modern equivalent of the Aztec disease due to difficulties in interpreting sixteenth century diagnoses. It might actually have been some kind of skin disease, and the plant might be useful. *P. zeylanica* is used in India, Japan, Ghana, and the Phillipines for skin diseases (Ayensu 1978: 216–218).

Plumeria rubra. Kaempferol and plumieride are purgative, which would be febrifuges by Aztec etiology. Plumieride is used in the West Indies (Ayensu 1981: 43). *P. attenuata* is used in Brazil as a purgative and against fevers; *P. rubia* is used as a laxative in Yucatan (Morton 1981: 677–678; von Reis Altschul 1973: 224).

Polyanthes tuberosa. This plant should not produce the desired effects.

Polygonum hydropiper. The astringent and hemostatic effects of tannins would be useful for sore or injured feet. It is used as a wound remedy in China (von Reis Altschul 1973: 62).

Prosopis juliflora. Various tribes used *Prosopis* sp. for sore eyes, so the gum can be expected to be a demulcent similar to acacia (Vogel 1970: 322; Moerman 1986: 369). It is used similarly in Mexico and Curazao (Morton 1981: 344–345).

Prunus capulli. Due to its tannin and pectin content it is effective against diarrhea (Orellana 1987: 231). It was used in folk medicine in Guatemala and Yucatan for this purpose (Morton 1981: 265). Extracts were shown to have anti-inflammatory effects in the laboratory (Jiu 1966).

Psidium guajava. The tannin content of this plant makes it effective against diarrhea. It was used for this purpose in a number of countries in Central America, the West Indies, and Africa (Ayensu 1981: 199; Morton 1981: 629–630). Extracts have shown antibacterial and antimicrobial activity (Ayensu 1978: 136; Jiménez et al. 1979).

Quercus sp. Tannins are styptic, astringent, and effective against diarrhea (Spoerke 1980: 129). A number of *Quercus* species were used by various North American tribes and as folk medicine in Cuba and Mexico for antidiarrheal remedies (Moerman 1986: 389–395; Morton 1977: 135).

Rhamnus serrata. Anthraquinone glycosides are purgatives, directly stimulating the gastrointestinal tract (Spoerke 1980: 36, 46, 50).

Rumex mexicana. Chrysophanic acid is an effective purgative, and its tannins account for its astringency. Morton (1981: 174–175) pointed out that overdoses of *R. crispus* produce excessive urination.

Salix lasiopelis. Salicin has analgesic and antirheumatic properties. It reduces fever by interfering with signals to the temperature control center in the hypothalamus. It dilates blood vessels, helping to reduce temperature. Its interference with prostaglandin biosynthesis is anti-inflammatory (der Marderosian and Liberti 1988: 37). It has been used for fever and against diarrhea by the Mendocino Indians and in Venezuela and Guatemala (Moerman 1986: 429; Morton 1981: 132).

Sambucus mexicana. Sambucus species contain a purgative resin. Bark, berries, and leaves have diuretic, laxative, and diaphoretic powers (Spoerke 1980: 68–69). Their flavonol content is responsible for the diuretic action. The plant has been used as a diuretic in Guatemala, Mexico, and Venezuela and as a purgative in Colombia and Venezuela (Mellen 1974; Morton 1981: 880–881; von Reis and Lipp 1982: 295).

Schoenocaulon coulterii. Veratrum alkaloids are extremely irritating to mucous membranes. This plant was formerly used medicinally as a topical anodyne counterirritant. Inserted in the nose as a powder, it would cause violent sneezing leading to nosebleed. This would be a remedy for headache according to Aztec etiology. These compounds are quite toxic and effective in killing mice. Extracts of *S. officinale* showed strong specific toxic action towards houseflies.

Smilax aristolochiaefolia. The sapogenin content of this plant produces diuretic and diaphoretic effects (Spoerke 1980: 154–155; Hegnauer 1962–1977: vol. 2, 346). It has been used for these purposes in folk medicine all over the world (Morton 1981: 82–85; Keys 1976: 59; Ayensu 1981: 250).

Stevia salicifolia (Sahagún 1950–1969: book 10, 152; book 11, 167, 197; Hernández 1942–1946: vol. 1, 56). Saponins are diuretic and validate Aztec use as a fever remedy. There are reports of its folk use for fevers in Mexico (von Reis Altschul 1973: 298).

Tagetes erecta. Glycosides of quercetin are diaphoretic. This plant is used as a purgative in Cuba (Morton 1981: 969–970).

Tagetes lucida. Flavonoids, particularly quercetin glycosides, are diuretic. Neher (1968) reported that tea made from *T. lucida* has a strong diuretic effect. By Aztec etiology, fever, fear, dementia and the aftereffect of lightning were all phlegm diseases; a diuretic would have been an appropriate remedy.

Talauma mexicana. Quercetin would account for this plant's diuretic action and use for fevers by the Aztecs. Extracts of produced vasoconstriction, arterial hypotension, and increase in the force of contractions and muscule tone of the heart (Lozoya and Lozoya 1982: 111–129). Extracts also relax the bladder and intestinal tissue of rabbits and contract the intestinal tissue of guinea pig (Lozoya et al. 1978).

Teloxys ambrosioides (Chenopodium ambrosioides). Ascaridol is antiseptic in dilutions of 1:200. It is an effective anthelmintic and produces a decrease in gastric motility leading to constipation (Spoerke 1980: 138–139; Lozoya and Lozoya 1982: 31–34); Bye and Linares 1987). It is widely used in folk medicine by North American Indians and in the West Indies, Mexico, and Central America (Moerman 1986: 114; Morton 1981: 176–177; von Reis Altschul 1973: 63; Ayensu 1981: 75).

Thalictrum hernandezii. Berberine is used for the relief of inflamed eyes, as discussed for *Argemone mexicana*.

Theobroma cacao. Theobromine is a diuretic, stimulant, and smooth-muscle relaxant and is used pharmaceutically for these purposes (Morton 1981: 556–557). It is not clear if mixing rubber with chocolate would stop diarrhea. The 2-phenylethyl amine can have several side effects if taken in large quantities. It has an amphetamine-like stimulant action and has been reported to cause hallucinations (Orellana 1987: 247).

Turbina corymbosa (Rivea corymbosa) (Sahagún 1950–1969: book 11, 129, 165, 170; Hernández 1959: vol. 2, 73). Ergine alkaloids and other derivatives found in this plant are clearly hallucinogenic (Schultes and Hofmann 1980: 240–257). The presence of ergonovine makes this plant oxytocic (Farnsworth et al. 1975; Lozoya and Lozoya 1982: 231–247). Atropine, which penetrates the skin and produces an anodyne effect, would make a topical application of *Datura* for gout logical. It is not clear, however, that the compounds in *Turbina*, which is also hallucinogenic, would have the same effect. Some reports cite the folk use of this plant as a local anesthetic and as an analgesic poultice for rheumatism (Díaz 1979; Morton 1981: 703–704).

Notes

CHAPTER 1

1. For a more extensive description of the evolution of civilization in Mesoamerica, see Coe (1984) or Adams (1977).
2. For a fuller description of the social, economic, and political organization of the Aztecs, see Berdan (1982) and Soustelle (1970).
3. This work is properly titled *Libellus de Medicinalibus Indorum Herbis*, "Book of the Medicinal Herbs of the Indians," but is most often referred to as the *Badianus Codex*, after the name of the translator.
4. Abdera was a town in Thrace whose inhabitants had a reputation for folly and stupidity (de la Cruz 1940: 310).
5. *Protomedicos* were a group of physicians empowered to license other doctors and to supervise medical standards. Thus, Hernández was given a supervisory and superior rank over physicians in the New World as a symbol of the importance of his mission and to assist him in getting cooperation in his task.

CHAPTER 2

1. Karttunen and Lockhart (1987: 31) disagreed with this etymology, but the analogy is still illuminating and fits the Aztec ethos.
2. Bierhorst (1985) published a radically different translation of these poems. Both Léon Portilla (1986) and the present author (Ortiz de Montellano n.d.) disagree fundamentally with his version. The translation given here is accepted by most Nahuatl scholars.

CHAPTER 3

1. The daily corn requirement is derived by modifying Cook and Borah
    (1979: 157), adding 300 cal/hr for a 5-hr workday to the requirement of
    the tlameme and taking into account the lesser needs of other mem-
    bers of the family.

9 hr sleep @ 53.6 cal/hr	482 cal
15 hr waking	
basal rate @ 59.4 cal/hr	891
digestion	140
5 hr work @ additional 300 cal/hr	1,500
10 hr light work @ 37 cal/hr	370
TOTAL	3,383

    However, the tlameme would not work every day. Assuming at a re-
    duced requirement of 2,068 cal/day and 290 days at 3,383 cal/day 75
    days off, the average daily requirement over a year would be:

    $$3,383 \times 290 + 2,068 \times 75)/365 = 1,136,170/365 = 3,113 \text{ cal/day}$$

    The daily requirement for a family of five is then:

adult male	3,113 cal
adult female	2,107
3 children @ 1,270 cal/day	3,810
TOTAL	9,030 cal/day

    $$\text{amount of corn} = (9,030 \text{ cal/day})/(4,000 \text{ cal/kg}) = 2.26 \text{ kg/day}$$

2. If we assume that the lakes form a circle with an area of 1,000 km², its
    radius would be 17.8 km. An effective economic area that extends 5
    leagues (28 km) from the shoreline has a radius of 17.8 + 28 = 45.8 km.
    For 15 leagues (84 km), the radius is 17.8 + 84 = 101.8 km. The effective
    economic areas are then (subtracting the lake area):

    a) 5 leagues
    $$\pi \times 45.8^2 - 1,000 = 5.6 \times 10^3 \text{ km}^2$$
    b) 15 leagues
    $$\pi \times 101.8^2 - 1,000 = 3.16 \times 10^4 \text{ km}^2$$

3. "Despite general agreement that hyperalimentation ought to work
    and the spending of billions of dollars on postoperative nutritional
    support there is only limited objective evidence for its efficacy in
    terms of reduced mortality and morbidity, and even fewer data show-
    ing effects at the tissue level. There are, for example, no reports of in-
    travenous feeding causing positive nitrogen balance during the "flow"
    phase in severely traumatized, burned, or even depleted cancer pa-
    tients (Rennie and Harrison 1984)."

CHAPTER 4

1. Green plants absorb radiation only in the visible spectrum (wavelengths of 400–700 nm), which makes up 43 percent of the total radiation incident on the earth's surface. The energy in each photon (unit quantity, or quantum, of light) is inversely proportional to its wavelength. For simplicity, the average visible energy can be expressed as if all its photons had a wavelength of 575 nm. Since chemical reactions are usually calculated in terms of mole quantities (Avogadro's number 6.02×10^{23} molecules), we will use a mole of photons, called an einstein, as our unit, rather than single photons. Applying the appropriate proportionality constant, we get $28,600/575 = 49.74$ kcal for the energy in an einstein of visible light. Since photosynthesis of the basic carbohydrate unit (CH_2O) involves the absorption of 8 einsteins, or $8 \times 49.74 = 398$ kcal of energy, for an output of 114 kcal, the maximum theoretical thermodynamic efficiency is:

    Thermodynamic efficiency = Energy out/Energy in = $114/398$
    = 0.286, or 28.6%

    However, this is for visible light only. The actual energy input includes all solar radiation, that is, $1/0.43 = 2.33$ times the visible portion. Since the denominator of the efficiency ratio is increased by a factor of 2.33, the overall efficiency is correspondingly reduced to $28.6/2.33 = 12.3\%$ (Bassham 1977).

2. The upper limit of the canopy effect has been estimated at 0.80 and of the respiration effect at 0.667 (Bassham 1977).

    Efficiency (land plants) = $0.123 \times 0.80 \times 0.667$
    = 0.066, or 6.6%

3. The annual solar radiation in the southwest United States averages 4,610 kcal/m² per day. The maximum theoretical yield in this case is calculated as follows:

    a) Energy converted to glucose = $0.066 \times 4,610 = 304.3$ kcal/m² per day
    b) Amount of the basic carbohydrate unit produced at 114 kcal/mole (6.023×10^{23} molecules = 1 mole) is then $304.3/114 = 2.67$ mole/m² per day
    c) The basic component of biomass is glucose, from which both starch and cellulose are made. Glucose ($C_6H_{12}O_6$) is composed of 6 basic carbohydrate units. Thus, the yield of glucose (cellulose) is $2.67/6 = 0.445$ mole/m² per day. At 162 g/mole, this is $162 \times 0.445 = 72.1$ g/m² per day.

4. C_4 plants use atmospheric CO_2 efficiently by concentrating it in the chloroplasts of specialized cells surrounding the leaf vascular bundles. By thus suppressing loss of CO_2 by respiration, C_4 plants convert

a greater amount of atmospheric CO_2 into carbohydrate than do C_3 plants, which do not possess this mechanism.

5. The minimum daily requirement for protein relative to body weight is 0.71 g/kg. If we assume all Aztecs to weigh 60 kg, the requirement would be 42.6 g/day per person. The area of Lake Texcoco in 1500 A.D. was approximately 560 km^2 or 56,000 ha. Populations were 300,000 (3 × 10^5) for Tenochtitlan and 1.2 million (1.2 × 10^6) for the Basin of Mexico.

    a) Protein need per year:
        Tenochtitlan: $3 \times 10^5 \times 42.6 \times 365 = 4.665 \times 10^9$ g
        At 10^6 g per metric ton, this equals 4,665 ton/yr.
        Basin of Mexico: $1.2 \times 10^6 \times 42.6 \times 365 = 1.866 \times 10^{10}$ g.
        This equals 18,660 ton/yr.
    b) Areas needed at an annual yield of 35 ton/ha:
        Tenochtitlan:
        Area = 4,665/35 = 133.3 ha
        Fraction of lake surface = 133.3/56,000 = 0.0024, or 0.24%
        Basin of Mexico:
        Area = 18,660/35 = 533.1 ha
        Fraction of lake surface = 533.1/56,000 = 0.0095, or 0.95%

6. The Otomí would be judged to fare even better today, since RDA levels have been decreasing as the body characteristics of non-Western peoples are taken into account. Relative to present RDA standards, the Otomí intake of calories was 78%, protein 113%, calcium 100%, and vitamin C 213%. The deficit of niacin decreases from 79% to 47%. The only nutrient deficiency evident both clinically and in dietary records was of riboflavin.

7. The number of insects in one percent (0.01 × 56,000 = 560 ha = 5.60 × 10^6 m^2) of the area of Lake Texcoco within a one-meter depth at 200,000 (2 × 10^5) insects/m^3 would be:

    $$5.60 \times 10^6 \times 2 \times 10^5 = 1.12 \times 10^{12}$$

    The weight of the insects, assuming a density of 0.25 g/cm^3 for dried insects 6 mm long, 2 mm wide, and 1 mm thick (volume = 6 × 2 × 1 = 12 mm^3 = 1.2 × 10^{-2} cm^3 would be:

    $$1.12 \times 10^{12} \times 1.2 \times 10^{-2} \times 0.25 = 3.36 \times 10^9 \text{ g}$$

    At 0.682 g of protein for each gram of insects, the protein derived from axayacatl in one percent of lake surface would be:

    $$3.36 \times 10^9 \times 0.682 = 2.29 \times 10^9 \text{ g, or } 2.29 \times 10^3 \text{ metric tons}$$

CHAPTER 6
1. López Austin (1988: vol. 1, 353) presented an extensive list of etiological factors in death and disease.

2. Andrews and Hassig (Ruiz de Alarcón 1984: 320) disagreed with Garibay's translation but agree that the name is onomatopoeic.

3. The text accompanying this plate reads: "These are the twenty characters or figures that they used in all their calculations, which they said had dominion over men as shown here, and accordingly to this they cured them, when someone got sick or certainly when a particular part of the body hurt. Bufeo [cipactli] had dominion over the liver. The rose over the breasts. Earthquake (*ollin*, the movement) over the tongue. The eagle over the right arm. The vulture on the right ear. The rabbit over the left ear. Flint on the teeth. The wind on the breath. The monkey over the left arm. The reed over the heart. The *mallinalli* on the intestines. The lizard over the womb of women. The tiger over the left foot. The snake on the male penis as a thing from which has come the origin of his ailment."

4. This presumes an etiology involving possession by supernatural spirits. An alternate etiology involving an excess of phlegm in the chest is discussed later in the chapter.

5. Tobacco was mentioned in many contexts in these incantations under a variety of names such as "the-nine-beaten-one." As stated earlier, picietl was considered an avenue of communication with the gods and sacred to the extent that a tobacco gourd, the *yetecomatl*, was a characteristic of priestly garb.

6. According to López Austin (1988: vol. 1, 168), the Aztecs thought that a major vein ran from the genitals to the heart through the navel. As a major distribution center for conduits to the whole body, the navel would be a natural site for a warming rub.

CHAPTER 7

1. Copal was used in many religious rituals and curing ceremonies. In Mesoamerica, smoke was considered a vehicle to reach other worlds. For example, objects needed by the dead in the underworld were burned so that the smoke would transmit them (Sahagún 1956: vol. 1, 295). Smoke was also a Mayan vehicle for communicating with gods (Schele and Miller 1986: 43, 176, 178).

2. Crude extracts of plants may contain unidentified substances so that their actions are less conclusive evidence than the actions of pure compounds. The interaction of several components in mixtures may also have quite different effects than would be expected of single compounds acting alone.

3. In order to reduce the number of references, only additional citations concerning the botanical identity, chemical components, or physiological actions not included in the original papers will be cited here. The original work (Ortiz de Montellano 1974/75; 1975) should be consulted for the full references.

CHAPTER 8

1. Sahagún (1956: vol. 1, 121; 1950–1969: book 2, 115) mentioned the use of yauhtli as a stupefacient for prospective sacrificial victims. I doubt its effectiveness since it lacks the necessary chemicals, and its mode of administration, powder thrown at the person's face, would not deliver an adequate dose.

2. Some of these characteristics are: 1) He was born on a day 1-rain, that is, was predestined. 2) He is a priest of the Rain God. 3) He visits Tlalocan, which has an abundance of rain and food. 4) He regulates the amount of rainfall and thus whether the crop will be good or poor.

Bibliography

ABBREVIATIONS

IMEPLAM	Instituto Mexicano para el Estudio de Plantas Medicinales
IMSS	Instituto Mexicano del Seguro Social
INAH	Instituto Nacional de Antropología e Historia
INCAP	Instituto de Nutricíon Centro America y Panamá
UNAM	Universidad Nacional Autónoma de México

Ackernecht, E. H. 1965. *History and Geography of the Most Important Diseases*. New York: Hafner.

Acosta, Fr. J. de. 1962. *Historia Natural y Moral de las Indias*, ed. E. O'Gorman. México: Fondo de Cultura Económica (orig. 1590).

Adams, R.E.W. 1977. *Prehistoric Mesoamerica*. Boston: Little, Brown.

Aguilera, C. n.d. *Flora y Fauna Mexicana. Mitología y Tradiciones*. México: Editorial Everest.

Aguirre Beltrán, G. 1963. *Medicina y Magia*. México: Inst. Nacional Indigenista.

Alarcón-Segovia, D. 1980. Descriptions of therapeutic arthrocentesis and of synovial fluid in a Nahuatl text from pre-Hispanic Mexico. *Annals of Rheumatic Diseases* 39: 291–293.

Alcorn, J. B. 1984. Development policy, forests, and peasant farms: Reflections on Huastec-managed forests' contributions to commercial production and resource conservation. *Economic Botany* 38: 389–406.

Alva Ixtlilxochitl, F. de. 1975. *Obras Históricas*, ed. E. O'Gorman. México: UNAM.

Alvarez, L. 1976. Breve estudio de las plantas medicinales de Hueyapan, Morelos. In *Estudios sobre Etnobotánica y Antropología Medica*, ed. C. Viesca Treviño, 85–111. México: IMEPLAM.

Alvarez, M. L., M. T. Araneda, E. Figueróa, and S. Osorio. 1983. Tratamiento de enfermedades en una poblacion rural: ¿Vigencia de elementos hispanicos? *Social Science & Medicine* 17: 471–474.

American Herbal Pharmacology Delegation. 1975. *Herbal Pharmacology in the People's Republic of China*. Washington, D.C.: National Academy of Science.

Ancona, L. 1933. El ahuautle de Texcoco. *Anales Instituto de Biología* (México) 4: 51–69.

———. 1934. Los gusanitos de Maguey. *Anales Instituto de Biologia* (México) 5: 193–200.

Anderson, A.J.O. and C. E. Dibble. 1982. *Introductions and Indices. Florentine Codex. General History of New Spain*. Salt Lake City: Univ. of Utah Press.

Anderson, E. N. 1988. *The Food of China*. New Haven: Yale Univ. Press.

Anderson, R. K., J. Calvo, G. Serrano, and G. C. Payne. 1946. A study of the nutritional status and food habits of Otomi Indians in the Mezquital Valley of Mexico. *American Journal of Public Health* 36: 883–903.

Angell, M. 1985. Disease as a reflection of the psyche. *New England Journal of Medicine* 312: 1570–1572.

Anonymous. 1886–1892. Códice de Tlatelolco. In *Nueva Coleccíon de Documentos Inéditos para la Historia de México*, 5 vols., ed. J. García Icazbalceta, 241–271. México: Díaz de León (orig. 1572).

———. 1975. Leyenda de los soles. In *Códice Chimalpopoca. Anales de Cuauhtitlan y Leyenda de los Soles*, trans. P. F. Velázquez, 119–142. México: UNAM (orig. 1558).

———. 1981. A touch of glass. *Science 81* (September): 79–80.

———. 1985. Emotion and immunity. *Lancet ii* (July 20): 133–134.

Anonymous Conqueror. 1972. *Narrative of Some Things of New Spain and of the Great City of Temestitlan, Mexico*. Boston: Milford House (orig. 1556).

Araya, H., M. Flores, and G. Arroyave. 1981. Nutritive value of basic foods and common dishes of the Guatemalan rural populations. A theoretical approach. *Ecology of Food and Nutrition* 11: 171–176.

Ashburn, P. M. 1947. *The Ranks of Death: A Medical History of the Conquest of America*. New York: Coward McCann.

Ayensu, F. S. 1978. *Medicinal Plants of West Africa*. Algonac, MI: Reference Publications.

———. 1981. *Medicinal Plants of the West Indies*. Algonac, MI: Reference Publications.

Baker, B. J., and G. J. Armelagos. 1988. The origin and antiquity of syphilis. *Current Anthropology* 29: 703–737.

Balzer, M. M. 1983. Doctors or deceivers? The Siberian Khanty shaman and Soviet medicine. In *The Anthropology of Medicine*, ed. L. Romanucci-Ross, D. E. Moerman, and L. R. Tancredi, 54–76. New York: Bergin.

Bartrop, R. W., E. Luckhurst, L. Lazarus, L. G. Kiloh, and R. Penny. 1977. Depressed lymphocyte function after bereavement. *Lancet i* (April 16): 834–836.

Bassham, J. A. 1977. Increasing crop production through more controlled photosynthesis. *Science* 197: 630–638.

Bastien, J. 1983. Personal communication.

———. 1985. Qollahuaya-Andean body concepts: A topographic-hydraulic model of physiology. *American Anthropologist* 87: 595–611.

Beaton, G. H. and L. D. Swiss. 1974. Evaluation of the nutritional quality of food: Prediction of "desirable" and "safe" protein:calorie ratios. *American Journal of Clinical Nutrition* 27: 485–504.

Becker, R., E. L. Wheeler, K. Lorenz, A. E. Stafford, O. K. Grosjean, A. A. Betschart, and R. M. Saunders. 1981. A compositional study of amaranth grain. *Journal of Food Science* 46: 1176–1180.

Berdan, F. F. 1982. *The Aztecs of Central Mexico*. New York: Holt, Rinehart and Winston.

Berdan, F. F. and J. de Durand-Forest, eds. 1980. *Matricula de Tributos*. Graz, Austria: Akademische Druck (facsim.).

Berkow, R., ed. 1977. *The Merck Manual*, 13th ed. Rahway, NJ: Merck and Co.

Berlin, B., D. E. Breedlove, and P. H. Raven. 1966. Folk taxonomies and biological classification. *Science* 154: 273–275.

———. 1974. *Principles of Tzeltzal Plant Classification*. New York: Academic Press.

Bierhorst, J., trans. 1985. *Cantares Mexicanos. Songs of the Aztecs*. Stanford: Stanford Univ. Press.

Bock, P. K. 1980. Tepoztlan reconsidered. *Journal of Latinamerican Lore* 6 (1): 129–150.

Boerhaave, H. 1755. *Aphorisms*. London: W. Innys, J. Richardson, C. Hitch, and L. Harves (orig. 1709).

Bolton, R. 1981. Susto, hostility and hypoglycemia. *Ethnology* 19: 261–276.

Borah, W. 1980. Cinco siglos de produccion y consumo de alimentos en el Mexico Central. Discurso de Recepcion como Corresponsal de La Academia Mexicana de Historia. Mexico (mimeo).

Borah, W. and S. F. Cook. 1958. *Price Trends of Some Basic Commodities in Central Mexico, 1531–1570*. Los Angeles: Univ. of California Press.

———. 1963. *The Aboriginal Population of Central Mexico on the Eve of the Spanish Conquest*. Los Angeles: Univ. of California Press.

Boston Women's Health Book Collective. 1984. *The New Our Bodies Ourselves*, 2d ed. New York: Simon & Schuster.

Boulos, L. 1983. *Medicinal Plants of North Africa*. Algonac, MI: Reference Publications.

Bresani, R. 1976. The role of small animal species in nutrition and food production. *Bulletin Pan American Health Organization* 10 (4): 293–300.

Brody, H. 1980. *Placebos and the Philosophy of Medicine*. Chicago: Univ. of Chicago Press.

Browner, C. H., B. R. Ortiz de Montellano, and A. J. Rubel. 1988. A methodology for cross-cultural ethnomedical research. *Current Anthropology* 29: 681–702.

Bryant, J. 1969. *Health and the Developing World*. Ithaca: Cornell Univ. Press.

Bustamante, M. E. 1975. Aspectos históricos y epidemiológicos del hambre en México. *Gaceta Medica de México* 109 (1): 23–43.

———. 1984. Saneamiento del medio en los pueblos prehispánicos. In *Historia General de la Medicina en México*, ed. A. López Austin and C. Viesca Treviño, vol. 1, 385–391. México: UNAM-Academia Nacional de Medicina.

Bye, R. 1981. Quelites-ethnoecology of edible greens—past, present, and future. *Journal of Ethnobiology* 1 (1): 109–123.

———. 1986. Personal communication.

Bye, R. and M. E. Linares. 1987. Usos pasados y presentes de algunas plantas medicinales encontradas on los mercados mexicanos. *America Indigena* 47: 199–230.

Caballero, Y. and F. Walls. 1970. Productos naturales del zoapatle. *Bol. Instituto Quimica (UNAM)* 22: 79–102.

Cáceres, A. 1984. In xochitl, in cuicatl: Hallucinogens and music in Mesoamerican Amerindian thought. Ph.D. dissertation. Indiana University, Bloomington.

Callen, E. O. 1973. Dietary patterns in Mexico between 6500 B.C. and A.D. 1580. In *Man and his Foods*, ed. C. E. Smith, 29–49. University: Univ. of Alabama Press.

Candelaria, A. 1976. Personal communication.

Carpenter, C.C.J. 1982. Oral rehydration. Is it as good as parenteral therapy? *New England Journal of Medicine* 306: 1103–1104.

Carrasco, D. 1982. *Quetzalcoatl and the Irony of Empire*. Chicago: Univ. of Chicago Press.

Carrera Stampa, M. 1949. The evolution of weights and measures in New Spain. *Hispanic American Historical Review* 29: 2–24.

Casillas, L. E. 1978. El uso de recursos médicos en el hogar: Estudio de familias urbanas y suburbanas en la ciudad de México. In *Estudio sobre Etnobotánica y Antropología Medica*, ed. C. Viesca Treviño, vol. 3, 95–114. México: IMEPLAM.

Caso, A. 1958. *The Aztecs People of the Sun*. Norman: Univ. of Oklahoma Press.

———. 1967. *Los Calendarios Prehispánicos*. México: UNAM.

———. 1971. ¿Religíon o religiones Mesoamericanas? *Verhandlungen des XXXIII Internationalen Amerikanisten Kongresses* (Stuttgart-Munchen 1968) 3: 189–200.

Castiglioni, A. 1947. *Encantamiento y Magia*. Mexico: Fondo de Cultura Economica.

Castillo, V. M. 1972. *Estructura Económica de la Sociedad Mexica*. México: UNAM.

Celsus. 1935–1938. *De Medicina*, 3 vols., trans. W. G. Spencer. Cambridge: Harvard Univ. Press.

Chaney, M. S. and M. L. Ross. 1971. *Nutrition*, 8th ed. New York: Houghton Mifflin.

Chen, A. 1987. Unraveling another Maya mystery. *Discover* (June): 40–49.

Chirife, J., L. Herzage, A. Joseph, and E. S. Kohn. 1983. In-vitro study of bacterial growth inhibition in concentrated sugar solutions—microbiological basis for the use of sugar in the treatment of infected wounds. *Antimicrobial Agents and Chemotherapy* 23: 766–773.

Christian, W. A. 1981. *Local Religion in Sixteenth-Century Spain*. Princeton: Princeton Univ. Press.

Ciferri, O. 1981. Let them eat algae. *New Scientist* (September 24): 810–812.

Claus, E. P., V. E. Tyler and L. R. Brady. 1970. *Pharmacognosy*. 6th ed. Philadelphia: Lea and Febiger.

Clavijero, F. J. 1971. *Historia Antigua de México*. Mexico: Porrúa.

Clement-Brutto, S. n.d. Symbolic clusters and cognition: A Salvadoran example. Typescript.

Codex Borbonicus. 1974. Ed. K. A. Nowotny and J. de Durand-Forest. Graz: Akademische Druck (facsim.).

Codex Borgia. 1963. México: Fondo de Cultura Economica (facsim.).

Codex Boturini. 1964. In *Antiguedades de México*, 4 vols. Mexico: Secretaría de Hacienda (facsim.).

Codex Fejervary-Mayer. 1964. In *Antiguedades de México*, 4 vols. México: Secretaría de Hacienda (facsim.).

Codex Laud. 1964. In *Antiguedades de México*, 4 vols. Mexico: Secretaría de Hacienda (facsim.).

Codex Magliabechiano. 1983. Ed. Z. Nuttall. Berkeley: Univ. of California Press (reprint of 1903 ed.).

Codex Vaticanus Latinus 3738 (*Codex Ríos*). 1964. In *Antiguedades de México*, 4 vols. México: Secretaría de Hacienda (facsim.).

Coe, M. D. 1964. The chinampas of Mexico. *Scientific American* 211: 90–98.

———. 1975. Death and the ancient Maya. In *Death and the Afterlife in Pre-Columbian America*, ed. E. P. Benson, 87–104. Washington, D.C.: Dumbarton Oaks.

———. 1981. Religion and the rise of Mesoamerican states. In *The Transition to Statehood in the New World*, ed. G. D. Jones and R. R. Kautz, 157–171. Cambridge: Cambridge Univ. Press.

———. 1984. *Mexico*, 3d ed. New York: Thames and Hudson.

Coe, S. 1985. Aztec cuisine. Part II. *Petit Propos Culinaire* 20: 44–59.

Cole, J. N. 1979. *Amaranth*. Emmaus, PA: Rodale Press.

Colson, A. B. and C. de Armellada. 1983. An Amerindian derivation for Latin American creole illnesses and their treatment. *Social Science & Medicine* 17: 1229–1248.

Comas, J. 1954. Influencia indígena en la medicina hipocrática en la Nueva España del siglo XVI. *América Indígena* 14: 327–361.

———. 1964. Un caso de aculturacíon farmacológica en la Nueva España del siglo XVI: El "Tesoro de Medicina" de Gregorio López. *Anales de Antropología* 1: 145–173.

———. 1968. La Medicina aborigen mexicana en la obra de Fray Agustín de Vetancurt. *Anales de Antropología* 5: 129–162.

———. 1971. Influencia de la farmacopea y terapéutica indígenas en Nueva España en la obra de Juan de Barrios. *Anales de Antropología* 8: 125–150.

Consumer and Food Economics Research Division. 1963. *Handbook of the Nutritional Content of Foods*. Washington, D.C.: U.S. Department of Agriculture.

Cook, S. F. 1946. The incidence and significance of disease among the Aztecs and related tribes. *Hispanic American Historical Review* 26: 320–335.

———. 1949. *The Historical Demography of the Teotlalpan*. Los Angeles: Univ. California Press.

Cook, S. F. and W. Borah. 1979. Indian food production and consumption in Central Mexico before and after the Conquest (1500–1650). In *Essays in Population History: Mexico and California*, vol. 3, 129–176. Los Angeles: Univ. of California Press.

Cook, S. F. and L. B. Simpson. 1948. *The Population of Central Mexico in the Sixteenth Century*. Los Angeles: Univ. of California Press.

Cook de Leonard, C. 1966. Roberto Weitlaner y los graniceros. In *Summa Anthropologica: En Homenaje a Roberto Weitlaner*, ed. A. Pompa y Pompa, 291–298. México: INAH.

Cooper Clark, J. 1938. *Codex Mendoza*, 3 vols. London: Waterlow and Sons.

Cope, Z. 1958. The treatment of wounds through the ages. *Medical History* 2: 163–174.

Cortés, H. 1971. *Letters from Mexico*, ed. and trans. A. R. Padgett. New York: Grossman (orig. 1522–1525).

Cortés, J. 1979. La medicina tradicional en la Sierra Mazateca. *Actes du XLIIe Congrès International des Americanistes*. (Paris 1976) 6: 349–356.

Cosminsky, S. 1976. The evil eye in a Quiche community. In *The Evil Eye*, ed. C. Maloney, 163–174. New York: Columbia Univ. Press.

———. 1980. Medical pluralism on a Guatemalan plantation. *Social Science & Medicine* 14B: 267–278.

Coury, C. 1969. *La Médecine de l'Amérique Précolombienne*. Paris: Roger Dacosta.

Cox, G. W. 1978. The ecology of famine: An overview. *Ecology of Food and Nutrition* 6: 207–220.

Crandon, L. 1983. Why susto? *Ethnology* 22: 153–167.

Cravioto, R. O. 1951. Valor nutritivo de los alimentos mexicanos. *Ciencia (México)* 11: 9–17.

Cravioto, R. O., Y. O. Cravioto, H. G. Massieu, and G. J. Guzmán. 1955. El pozol, forma indígena de consumir el maíz en el sureste de México y su aporte de nutrientes a la dieta. *Ciencia (México)* 15: 27–30.

Cravioto, R. O., E. E. Lockhart, R. K. Anderson, F. de P. Miranda, and R. S. Harris. 1945. Composition of typical Mexican foods. *The Journal of Nutrition* 29: 317–329.

Cravioto, R. O., G. Massieu, and J. Guzmán. 1951. Composicíon de alimentos mexicanos. *Ciencia (México)* 11: 129–155.

Crosby, A. W. 1972. *The Columbian Exchange. Biological and Cultural Consequences of 1492.* Westport, CT: Greenwood Press.

Davidson, J. L. and B. R. Ortiz de Montellano. 1983. The antibacterial properties of an Aztec wound remedy. *Journal of Ethnopharmacology* 8: 149–161.

Dawson, T. J. 1977. Kangaroos. *Scientific American* (August): 78–89.

Deevey, E. S. 1957. Limnologic studies in Middle America. *Transactions of the Connecticut Academy of Arts and Sciences* 39: 213–328.

de la Cruz, M. 1940. *The Badianus Manuscript (Codex Barberini, Latin 241)*, trans. E. W. Emmart. Baltimore: Johns Hopkins Univ. Press (orig. 1552).

———. 1964. *Libellus de Medicinalibus Indorum Herbis*, ed. E. C. del Pozo. Mexico: IMSS (orig. 1552).

de las Casas, Fr. G. 1967. *Apologética Historia Sumaria*, 2 vols., ed. E. O'Gorman. México: UNAM.

de la Serna, J. 1953. *Tratado de las Idolatrías, Supersticiones, Dioses, Ritos, Hechicerías y Otras Costumbres Gentílicas de las Razas Aborigenes de México.* México: Ediciones Fuente Cultural (orig. 1656).

del Pozo, E. C. 1964. Valor médico y documental del manuscrito. In *Libellus de Medicinalibus Indorum Herbis*, ed. E. C. del Pozo, 329–343. México: IMSS.

———. 1967. Empiricism and magic in Aztec pharmacology. In *Ethnopharmacological Search for Psychoactive Drugs*, ed. D. E. Efron, 59–76. Washington, D.C.: Government Printing Office.

Derbez, J. E. Pardo, and E. C. del Pozo. 1945. El cihuapatli, activador de la motilidad uterina. *Bol. Instituto Medico Biológico* 3: 127–139.

der Marderosian, A. and L. Liberti. 1988. *Natural Product Medicine.* Philadelphia: George F. Stickley.

Díaz, J. L. 1975. Etnofarmacología de algunos psicotrópicos vegetales de México. In *Etnofarmacología de Plantas Alucinógenas Latinoamericanas*, ed. J. L. Díaz, 35–201. Mexico: Centro Mexicano de Estudios en Farmacodependencia.

———. 1976a. Algunas plantas mexicanas con efecto sobre el sistema nervioso. In *Estado Actual del Conocimiento en Plantas Medicinales Mexicanas*, ed. X. Lozoya, 109–130. México: IMEPLAM.

———. 1976b. *Usos de las Plantas Medicinales de México.* México: IMEPLAM.

———. 1979. Ethnopharmacology and taxonomy of mexican psychodisleptic plants. *Journal of Psychedelic Drugs* 11: 71–101.

Díaz del Castillo, B. 1968. *Historia Verdadera de la Conquista de la Nueva España*. México: Porrúa (orig. 1568).

Dibble, C. E., ed. and trans. 1963. *Historia de la Nación Mexicana. Codice de 1576*. Madrid: Ediciones José Porrúa Turanzas.

――――. 1978. The conquest through Aztec eyes. *41st Annual Frederick Williams Reynolds Lecture*. Salt Lake City: Univ. of Utah Press.

Dick, H. 1946. A survey of astrological medicine in the age of science. *Journal of the History of Medicine* 1: 300–315, 419–433.

d'Olwer, N. and H. F. Cline. 1973. Sahagún and his work. In *Handbook of Middle American Indians*, ed. H. F. Cline, vol. 13, 187–207. Austin: Univ. of Texas Press.

Domínguez, X. A. and J. B. Alcorn. 1985. Screening of medicinal plants used by the Huaxtec Mayas of northeastern Mexico. *Journal of Ethnopharmacology* 13: 139–156.

Domínguez, X. A., R. Franco, and Y. Díaz-Vivecos. 1978. Mexican medicinal plants XXXIV. Rotenoids and a fluorescent compound from *Eysenhardtia polystachia*. *Revista Latinoamericana de Química* 9: 209–211.

Dornstreich, M. D. and G.E.B. Morren. 1974. Does New Guinea cannibalism have nutritive value? *Human Ecology* 2: 1–12.

Downs, P. 1984. Agriculture: Learning from ecology. *Technology Review* (July): 70–71.

Dubos, R. 1975. *Man Adapting*. New Haven: Yale Univ. Press.

Dunn, F. L. 1965. On the antiquity of malaria in the Western Hemisphere. *Human Biology* 37: 385–393.

Durán. Fr. D. 1967. *Historia de los Indios de la Nueva España e Islas de la Tierra Firme*, ed. A. M. Garibay, 2 vols. México: Porrúa (orig. 1581).

――――. 1971. *Book of the Gods and Rites and the Ancient Calendar*, ed. and trans. F. Horcasitas and D. Heyden. Norman: Univ. of Oklahoma Press (orig. 1579).

Durand-Chastel, H. 1980. Production and use of Spirulina in Mexico. In *Algae Biomass*, ed. G. Shelef and C. J. Soeder, 51–64. New York: Elsevier.

Eliade, M. 1963. *Myth and Reality*. New York: Harper & Row.

Escobar, G. J., E. Salazar, and M. Chung. 1983. Beliefs regarding the etiology and treatment of infantile diarrhea in Lima, Peru. *Social Science & Medicine* 17: 1257–1269.

Esteyneffer, J. de. 1978. *Florilegio Medicinal*, ed. M. C. Anzures, 2 vols. México: Academia Nacional de Medicina (orig. 1712).

Farb, P. and G. Armelagos. 1980. The food connection. *Natural History* 89 (9): 26–30.

Farnsworth, N. R. and A. S. Bingel. 1977. Problems and prospects of discovering new drugs from higher plants by pharmacological screening. In *New Natural Products and Plant Drugs with Pharmacological, Biological, or Therapeutic Activity*, ed. H. Wagner and P. Wolff, 1–22. Berlin: Springer Verlag.

Farnsworth, N. R., A. S. Bingel, G. A. Cordell, F. A. Crane, and H. S. Fong.

1975. Potential value of plants as sources of new antifertility agents. *Journal of Pharmaceutical Sciences* 64: 535–598, 717–754.

Fastlicht, S. 1971. *La Odontologia en el Mexico Prehispanico*. Mexico: Edimex.

Faulhaber, J. 1965. La poblacíon de Tlatilco, México, caracterizada por sus entierros. In *Homenaje a Juan Comas*, 83–121. México: UNAM.

Felger, R. S. and M. B. Moser. 1976. Seri Indian food plants: Desert subsistence without agriculture. *Ecology of Food and Nutrition* 5: 13–27.

Felker, P. and R. S. Bandurski. 1979. Uses and potential uses of leguminous trees for minimal energy input agriculture. *Economic Botany* 33: 172–184.

Fernández del Castillo, F. 1942. Algunos ejemplos de observacíon clínica en la medicina mágica azteca. *Revista Facultad de Medicina, UNAM* 12: 169–181.

Figueroa Marroquín, H. 1957. *Enfermedades de los Conquistadores*. San Salvador, El Salvador: Ministerio de Cultura.

Flores, M., E. Flores, B. García, and Y. Galarte. 1960. *Tabla de Composicion de Alimentos de Centro América y Panamá*. Guatemala City: INCAP.

Ford, K. C. 1975. *Las Yerbas de la Gente: A Study of Hispano-American Medicinal Plants*. Ann Arbor: Univ. of Michigan Museum of Anthropology.

Forrest, R. D. 1982. Development of wound therapy from the Dark Ages to the present. *Journal Royal Society of Medicine* 75: 268–273.

Foster, G. M. 1953. Relationships between Spanish and Spanish-American folk medicine. *Journal of American Folklore* 66: 201–217.

———. 1967. *Tzintzuntzan*. Boston: Little Brown.

———. 1978a. Hippocrates' Latin American legacy: "Hot" and "cold" in contemporary folk medicine. In *Colloquia in Anthropology*, ed. R. K. Witherington, vol. 2, 3–19. Dallas: Southern Methodist Univ. Press.

———. 1978b. Humoral pathology in Spain and Spanish America. In *Homenaje a Julio Caro Baroja*, ed. A. Carreira, J. A. Cid, M. Gutierrez Esteve, and R. Rubio, 357–370. Madrid: Centro de Investigaciones Sociologicas.

———. 1985. How to get well in Tzintzuntzan. *Social Science & Medicine* 21: 807–818.

———. 1986. On the origin of humoral medicine in Latin America. Paper presented at 85th Annual Meeting of the American Anthropological Association, 3–5 December, Philadelphia, PA.

———. 1987. On the origin of humoral medicine in Latin America. *Medical Anthropology Quarterly* 1: 355–393.

Foster, H. G. and B. G. Anderson. 1978. *Medical Anthropology*. New York: Alfred E. Knopf.

Fragoso, R. 1978. *Etnomedicina de los Actuales Matlazinca*. Master's thesis. México: Escuela Nacional de Antropología e Historia.

Furst, P. 1976. *Hallucinogens and Culture*. San Francisco: Chandler-Sharp.

———. 1978. Spirulina. *Human Nature* 1 (3): 60–65.

Garibay, A. M. 1943–1946. Paralipómenos de Sahagún. *Tlalocan* 1 (4): 307–313, 2 (2): 167–176, 2 (3): 235–254.

———. 1948. Relacíon breve de las fiestas de los dioses. *Tlalocan* 2 (4): 289–320.

———. 1949. Cuando el tecolote canta el indio muere. *Sociedad Folklórica de México, Anuario* 6: 383–399.

———. 1967. Códice Carolino. *Estudios de Cultura Nahuatl* 7: 11–58.

———. 1971. *Historia de la Literatura Nahuatl*, 2 vols. México: Porrúa.

———. 1973. *Teogonía e Historia de los Mexicanos: Tres Opúsculos del Siglo XVI*. México: Porrúa.

Garn, S. M. 1979. The noneconomic nature of eating people. *American Anthropologist* 81: 902–903.

Garn, S. M. and W. D. Block. 1970. The limited nutritional basis of cannibalism. *American Anthropologist* 72: 106.

Gibbs, R. D. 1974. *Chemotaxonomy of Flowering Plants*, 4 vols. Montreal: McGill–Queens Univ. Press.

Gibson, C. 1964. *The Aztecs under Spanish Rule*. Stanford: Stanford Univ. Press.

Gillen, J. 1948. Magical fright. *Psychiatry* 2: 387–400.

Gilman, A. G. et al., eds. 1980. *Goodman and Gilman's The Pharmacological Basis of Therapeutics*, 6th ed. New York: Macmillan.

Gómez de Orozco, F. 1945. Costumbres, fiestas, enterramientos, y diversas formas de proceder de los indios de Nueva España. *Tlalocan* 2(1): 37–63 (orig. 1553).

Gómez-Pompa, A. 1987. On Maya silviculture. *Mexican Studies/Estudios Mexicanos* 3: 1–17.

González, J. F. 1888. *La Flora de Nuevo León*. Nuevo León. México: Imprenta Católica.

Gonzalves de Lima, O. 1956. *El Maguey y el Pulque en los Códices Mexicanos*. México: Fondo de Cultura Económica.

Goodwin, J. S. and J. M. Goodwin. 1984. The tomato effect. *Journal of the American Medical Association* 251: 2387–2390.

Gordon, B. L. 1959. *Medieval and Renaissance Medicine*. New York: Philosophical Library.

Gordon, J. F. 1970. Algal proteins and the human diet. In *Proteins as Human Food*, ed. R. A. Lawrie, 328–347. Westport, CT: AVI Publishing Co.

Graham, J. S. 1976. The role of *curanderos* in the Mexican-American Folk system in West Texas. In *American Folk Medicine*, ed. W. D. Hand, 175–189. Berkeley: Univ. of California Press.

Greve, M. 1971. *A Modern Herbal*, 2 vols. New York: Dover (orig. 1931).

Guerra, F. 1966. Aztec medicine. *Medical History* 10: 315–338.

Guiteras Holmes, C. 1965. *Los Peligros del Alma. Visíon del Mundo de un Tzotzil*. México: Fondo de Cultura Economica.

Gupta, M. C., B. M. Gandhi, and B. N. Tandon. 1974. An unconventional legume—*Prosopis cineraria*. *The American Journal of Clinical Nutritional* 27: 1035–1036.

Hackett, C. J. 1963. On the origin of the human treponematoses (pinta, yaws, endemic syphilis, and venereal syphilis). *Bulletin of the World Health Organization* 29: 7–41.

Hahn, D. W., E. W. Erickson, M. T. Lai, and A. Probst. 1981. Antifertility activity of *Montanoa tomentosa* (zoapatle). *Contraception* 23: 133–140.

Harley, G. W. 1941. *Native African Medicine with Special Reference to Its Practice in the Mano Tribe of Liberia.* Cambridge: Harvard Univ. Press.

Harner, M. 1977. The ecological basis for Aztec sacrifice. *American Ethnologist* 4: 117–135.

Harris, M. 1978. The cannibal kingdom. In *Cannibals and Kings,* 147–166. New York: Random House.

———. 1979a. *Cultural Materialism. The Struggle for a Science of Culture.* New York: Random House.

———. 1979b. Our pound of flesh. *Natural History* 88 (8): 30–36.

Hartmann, F. 1973. *Paracelsus: Life and Prophecies.* Blauvelt, NY: Rudolph Steiner Publications.

Harwood, A. 1981. *Ethnicity and Health Care.* Cambridge: Harvard Univ. Press.

Hassig, R. 1981. The famine of one rabbit: Ecological causes and social consequences of a pre-Columbian calamity. *Journal of Anthropological Research* 37: 171–181.

———. 1985. *Trade, Tribute and Transportation. The Sixteenth Century Political Economy of the Valley of Mexico.* Norman: Univ. of Oklahoma Press.

Hegnauer, R. 1962–1977. *Chemotaxonomie der Pflantzen,* 7 vols. Basel: Birkhauser Verlag.

Hernández, F. 1942–1946. *Historia de las Plantas de la Nueva España,* 3 vols., ed. I. Ochoterena. México: Imprenta Universitaria (orig. 1577).

———. 1945. *Antiguedades de la Nueva España.* México: Editorial Pedro Robledo.

———. 1959. *Historia Natural de la Nueva España,* 2 vols. México: UNAM (orig 1577).

———. 1959–1985. *Obras Completas,* 6 vols. México: UNAM.

Hernández Rodríguez, R. 1962. Epidemias y calamidades en el Mexico prehispanico. *Anuario de Historia* 2: 21–35.

Herszage, L., J. R. Montenegro, and A. L. Joseph. 1980. Tratamiento de las heridas supuradas con azucar granulado comercial. *Boletines y Trabajos de la Sociedad Argentina de Cirujanos* 61: 315–330.

Hindley, K. 1979. Reviving the food of the Aztecs. *Science News* 116: 168–169.

Historia de los Mexicanos por sus Pinturas. 1973. In *Teogonía e Historia de los Mexicanos: Tres Opúsculos del Siglo XVI,* ed. A. M. Garibay, 23–87, México: Porrúa (orig. 1533).

Hobgood, J. J. 1971. The sanctuary of Chalma. *Verhandlungen des XXXVII Internationale Amerikanisten Kongress* (Stuttgart-Munchen 1968) 3: 247–263.

Holland, W. R. 1978. Mexican-American medical beliefs: Science or magic? In *Hispanic Culture and Health Care*, ed. R. A. Martinez, 79–119. St. Louis, MO: C. V. Mosby.

Horwith, B. 1985. A role for intercropping in modern agriculture. *BioScience* 35: 286–291.

Hubi, M. 1954. Algunas observaciones sobre folklore médico del departamento del Junín. *Perú Indígena* 13: 70–90.

Hudson, E. H. 1965. Treponematoses and man's social evolution. *American Anthropologist* 67: 885–901.

Hultkrantz, A. 1985. The shaman and the medicine man. *Social Science & Medicine* 20: 511–515.

Hunn, E. 1982. Did the Aztecs lack potential animal domesticates? *American Ethnologist* 9: 578–579.

Hurtado Vega, J. J. 1979. La "mollera caída" una subcategoría cognitiva de las enfermedades producidas por la ruptura del equilibrio mecánico del cuerpo. *Boletín Bibliográfico de la Antropología Mexicana* 41 (50): 139–148.

Ingham, J. M. 1986. *Mary, Michael, and Lucifer. Folk Catholicism in Central Mexico*. Austin: Univ. of Texas Press.

Instituto Médico Nacional. 1894–1907. *Datos para la Materia Médica Mexicana*, 4 vols. México: Secretaría de Fomento.

Jaén, Ma. T. 1977. Paleopatología en México. *Anales de Antropología* 14: 345–379.

Jaén, Ma. T. and S. López Alonzo. 1974. Algunas características físicas de la poblacíon prehispánica de México. In *Antropología Física, Epoca Prehispánica*, 113–165. Mexico: Secretaría Educacíon Pública–INAH.

Jarcho, S. 1964. Some observations on disease in prehistoric North America. *Bulletin of the History of Medicine* 38: 1–19.

Jenkins, K. D. 1964. Aje or Ni-in (the fat of the scale insect) painting medium and unguent. *Actas y Memorias XXXV Congreso Internacional de Americanistas*. (Mexico) 1: 625–636.

Jiménez, C. A., N. M. Hernández, A. M. López, and P. Kauri. 1979. A contribution to biological assessment of Cuban plants. *Revista Cubana Medicina Tropical* 31: 13–20.

Jiménez Moreno, W. 1971. Las religiones Mesoamericanas y el Cristianismo. In *Verhandlungen des XXXIII Internationalen Amerikanisten Kongresses* (Stuttgart-Munchen 1968) 3: 201–206.

Jiu, J. 1966. A survey of some medicinal plants of Mexico for selected biological activities. *Lloydia* 29: 250–259.

Joralemon, D. 1982. New world depopulation and the case of disease. *Journal of Anthropological Research* 38: 108–127.

Junod. H. 1927. *The Life of a South African Tribe*, vol. 1. New York: Macmillan.

Karim, S.M.M. 1972. *The Prostaglandins: Progress in Research*. New York: Wiley.

Karttunen, F. 1983. *An Analytical Dictionary of Nahuatl*. Austin: Univ. of Texas Press.

———. 1985. *Nahuatl and Maya in Contact with Spanish*. Austin: Department of Linguistics, Univ. of Texas.

Karttunen, F. and J. Lockhart, eds. and trans. 1987. *The Art of Nahuatl Speech. The Bancroft Dialogues*. Los Angeles: Univ. of California, Los Angeles, Latin American Center.

Katz, F. 1966. *Situación Social y Económica de los Aztecas Durante los Siglos XV y XVI*. México: UNAM.

Katz, S. H., M. L. Hediger, and L. A. Valleroy. 1974. Traditional maize processing techniques in the New World. *Science* 184: 765–773.

Kay, M. A. 1977a. The Florilegio Medicinal: Source of Southwest ethnomedicine. *Ethnohistory* 24: 251–259.

———. 1977b. Health and illness in a Mexican-American barrio. In *Ethnic Medicine in the Southwest*, ed. E. H. Spicer, 99–166. Tucson: Univ. of Arizona Press.

———. 1979. Lexemic change and semantic shift in disease names. *Culture, Medicine, and Psychiatry* 3: 73–94.

———. 1986. Personal communication.

Keen, B. 1971. *The Aztec Image in Western Thought*. New Brunswick: Rutgers Univ. Press.

Kendall, C. D., C. D. Foote, and R. Martorell. 1983. Anthropology, communications, and health: The mass media and health practices program in Honduras. *Human Organization* 42: 353–360.

Keys, J. D. 1976. *Chinese Herbs*. Rutland, VT: Charles F. Tuttle Co.

Kidwell, C. S. 1983. Aztec and European medicine in the New World, 1521–1600. In *The Anthropology of Medicine*, ed. L. Romanucci-Ross, D. E. Moerman, and L. R. Tancredi, 20–31. New York: J. F. Bergin.

Kimber, C. 1981. Personal communication.

Kirchhoff, P. 1943. Mesoamérica, sus limites geográficos, composicíon étnica y carácteres culturales. *Acta Americana* 1: 92–107.

Klor de Alva, J. J. 1988. Sahagún and the birth of modern ethnography: Representing, confessing, and inscribing the native other. In *The Work of Bernardino de Sahagún*, ed. J. J. Klor de Alva, H. B. Nicholson, and E. Quiñones Keber, 31–52. Albany: Institute for Mesoamerican Studies,, State Univ. of New York.

Knoke, B. 1984. Personal communication.

Kovar, A. 1970. The physical and biological environment of the Basin of Mexico. In *The Natural Environment, Contemporary Occupation and the Sixteenth Century Population of the Valley (Teotihuacan Valley Project Final Report)*, W. T. Sanders, A. Kovar, T. Charlton, and R. A. Diehl. University Park: Pennsylvania State Univ.

La Barre, W. 1970. Old and New World narcotics: A statistical question and an ethnographic reply. *Economic Botany* 2: 73–80.

Landgren, B. M., A. R. Aedo, K. Hagenfeldt, and E. Diczfaluzy. 1979. Clinical effects of orally administered extracts of *Montanoa tomentosa* in early human pregnancy. *American Journal of Obstetrics and Gynecology* 135: 480–484.

Latham, R. G., ed. and trans. 1848. *The Works of Thomas Sydenham*. London: The Sydenham Society (orig. 1666–1685).

León, N. 1910. *La Obstétrica en México*. México: Tipografía Vda. de F. Díaz de León.

León Portilla, M. 1961. *Los Antiguos Mexicanos*. México: Fondo de Cultura Económica.

———. 1963. *Aztec Thought and Culture*. Norman: Univ. of Oklahoma Press.

———. 1969. *Pre-Columbian Literatures of Mexico*. Norman: Univ. of Oklahoma Press.

———. 1986. ¿Una nueva interpretacíon de los "Cantares Mexicanos"? La obra de John Bierhorst. *Mexican Studies/Estudios Mexicanos* 2 (1): 129–144.

Levine, J. D., N. C. Gordon, and H. L. Fields. 1978. The mechanism of placebo analgesia. *Lancet ii* (September 23): 654–657.

Levine, S. D. et al. 1979. Zoapatanol and montanol novel oxepane diterpenoids, from the Mexican plant zoapatle (*Montanoa tomentosa*). *Journal American Chemical Society* 101: 3404–3405.

Levi-Strauss, C. 1967. *Structural Anthropology*. New York: Anchor Books.

Lewis, O. 1970. *Life in a Mexican Village. Tepoztlan Revisited*. Urbana: Univ. of Illinois Press.

Lewis, W. H. and M.P.F. Elvin-Lewis. 1977. *Medical Botany*. New York: John Wiley.

Littre, E., ed. 1846. *Oeuvres Complètes d'Hippocrate*. Paris: J. B. Ballière.

Litvak-King, J. 1971. *Cihuatlan y Tepecoacuilco: Provincias Tributarias de Mexico*. México: UNAM.

Logan, M. 1973. Digestive disorders and plant medicinals in highland Guatemala. *Anthropos* 65: 537–547.

López Austin, A. 1966. Los temacpalitotique. *Estudios de Cultura Nahuatl* 6: 97–117.

———. 1967. Cuarenta clases de magos del mundo nahuatl. *Estudios de Cultura Nahuatl* 7: 87–117.

———. 1969. *Augurios y Abusiones*. México: UNAM.

———. 1970a. Ideas etiológicas en la medicina nahuatl. *Anuario Indigenista* 30: 255–275.

———. 1970b. Conjuros magicos de los nahoas. *Revista de la Universidad de México* 24 (11): 1–16.

———. 1971a. *Textos de Medicina Nahuatl*. México: SEP-Setentas.

———. 1971b. De las plantas medicinales y de otras cosas medicinales. *Estudios de Cultura Nahuatl* 9: 125–230.

———. 1972. Textos acerca de las partes del cuerpo humano y de las enfermedades en los Primeros Memoriales de Sahagún. *Estudios de Cultura Nahuatl* 10: 129–153.

———. 1973. *Hombre Díos*. México: UNAM.

———. 1974a. Sahagún's work on the medicine of the ancient nahuas: Possibilities for study. In *Sixteenth Century Mexico. The Work of Sahagún*, ed. M. S. Edmonson, 205–224. Albuquerque: Univ. of New Mexico Press.

———. 1974b. The research method of Fray Bernardino de Sahagún: The questionnaires. In *Sixteenth Century Mexico. The Work of Sahagún*, ed. M. S. Edmonson, 111–149. Albuquerque: Univ. of New Mexico Press.

———. 1978. Discussion. In *Modern Medicine and Medical Anthropology in the United States—Mexico Border Population*, ed. B. Velimirovic, 201. Washington: Pan-American Health Organization.

———. 1980. *Cuerpo Humano e Ideología*, 2 vols. México: UNAM.

———. 1988. *Human Body and Ideology*, trans. T. Ortiz de Montellano and B. R. Ortiz de Montellano, 2 vols. Salt Lake City: Univ. of Utah Press.

López Austin, A. and C. Viesca Treviño, eds. 1984. *Historia General de la Medicina en México*, vol. 1. México: UNAM–Academia Nacional de Medicina.

Lozoya, M. 1978. Quauhtzapotl (*Annona cherimolia* L.). *Medicina Tradicional (Mexico)* 2(5): 1–4.

Lozoya, X. et al. 1978. Plantas mexicanas con uso popular: Su Validacíon experimental. *Medicina Tradicional (México)* 1 (3): 5–21.

Lozoya, X. and M. Lozoya. 1982. *Flora Medicinal de México*. México: IMSS.

Luborsky, L. B., B. Singer, and L. Luborsky. 1975. Comparative studies of psychotherapies: Is it true that "everyone has won and all must have prizes?" *Archives of General Psychiatry* 32: 995–1008.

McGregor, R. 1984. Los insectos en la dieta de los antiguos Mexicanos. In *Historia General de la Medicina en México*, ed. A. Lopez Austin and C. Viesca Treviño, vol. 1, 157–159. México: UNAM–Academia Nacional de Medicina.

McGuire, M. B. 1988. *Ritual Healing in Suburban America*. New Brunswick, NJ: Rutgers Univ. Press.

MacLachlan, C. M. and J. E. Rodriguez. 1980. *The Forging of the Cosmic Race*. Berkeley: Univ. of California Press.

McManus, J. J. 1969. Temperature regulation in the opossum *Didelphis marsupialis virginiana*. *Journal of Mammalogy* 50: 550–558.

McNeill, W. H. 1976. *Plagues and People*. New York: Doubleday.

Madsen, C. 1965. A study of change in Mexican folk medicine. *Middle American Research Institute Publication (Tulane University)* 25: 89–138.

Madsen, W. 1955. Hot and cold in the universe of San Francisco Tecospa. *Journal of American Folklore* 68: 123–138.

———. 1957. Christo-paganism. A study of Mexican religious syncretism. *Middle American Research Institute Publication (Tulane University)* 19: 105–180.

————. 1960. *The Virgin's Children*. New York: Greenwood Press.

————. 1964. *Mexican Americans of South Texas*. New York: Holt Rinehart & Winston.

————. 1967. Religious syncretism. In *Handbook of Middle American Indians*, ed. R. Wauchope, vol. 6, 369–391. Austin: Univ. of Texas Press.

————. 1969. The Nahua. In *Handbook of Middle American Indians*, ed. R. Wauchope, vol. 8, 602–637. Austin: Univ. of Texas Press.

Majno, G. 1975. *The Healing Hand: Man and Wound in the Ancient World.* Cambridge: Harvard Univ. Press.

————. 1983. Personal communication.

Mak, C. 1955. Mixtec medical beliefs and practices. *America Indígena* 19: 12–49.

Maloney, C. 1976. *The Evil Eye*. New York: Columbia Univ. Press.

Manheim, B. 1984. Personal communication.

Manuscrit Tovar. 1972. *Origins et Croyances des Indies du Mexique*, ed. J. Lafaye. Graz: Akademische Druck.

Martí, S. and C. Prokosh-Kurath. 1964. *Dances of Anahuac*. Chicago: Aldine.

Martin, F. W. and L. Telek. 1979. *Vegetables for the Hot, Humid Tropics. Part 6. Amaranthus and Celosia.* New Orleans: United States Department of Agriculture.

Martín del Campo, R. 1956. Productos biológicos del Valle de México. *Revista Mexicana de Estudios Antropológicos* 14 (1): 53–77.

Martínez, C. and H. W. Martin. 1966. Folk diseases among urban Mexican-Americans. *Journal American Medical Association* 196: 147–250.

Martínez, M. 1969. *Las Plantas Medicinales de México*. México: Botas.

Martínez Cortés, F. 1965. *Las Ideas en la Medicina Nahuatl*. Mexico: La Prensa Médica Mexicana.

Massieu, G., J. Guzman, R. O. Cravioto, and J. Calvo. 1951. Nutritive value of some primitive Mexican foods. *Journal of the American Dietetic Association* 27: 212–214.

Matos Moctezuma, E. 1987. Templo Mayor: History and interpretation. In *The Great Temple of Tenochtitlan*, ed. J. Broda, D. Carrasco, and E. Matos Moctezuma, 15–60. Berkeley: Univ. of California Press.

Mellen, G. A. 1974. El uso de las plantas medicinales en Guatemala. *Guatemala Indígena* 9: 100–179.

Mendieta, Fr. G de. 1971. *Historia Eclesiastica Indiana*. México: Porrúa (orig. 1596).

Meyer, R. J. and R. Haggerty. 1962. Streptococcal infections in families: Factors altering individual susceptibility. *Pediatrics* 29: 539–549.

Miller, A. G. 1975. *Codex Nuttall*. New York: Dover.

Moerman, D. E. 1979. Anthropology of symbolic healing. *Current Anthropology* 20: 59–80.

————. 1981. Edible symbols: The effectiveness of placebos. *Annals New York Academy of Sciences* 364: 256–268.

————. 1983. Physiology and symbols. The anthropological implications of the placebo effect. In *The Anthropology of Medicine*, ed. L. Romanucci-Ross, D. E. Moerman, and L. R. Tancredi, 156–167. New York: J. F. Bergin.

———. 1986. *Medicinal Plants of Native America*, 2 vols. Ann Arbor: Univ. of Michigan Museum of Anthropology.

Molina, Fr. A. de. 1970. *Vocabulario en Lengua Castellana y Mexicana y Mexicana y Castellana*. México: Porrúa (orig. 1571).

Molins Fábrega, N. 1954–1955. El Códice Mendocino y la economía de Tenochtitlan. *Revista Mexicana de Estudios Antropologicos* 14: 303–335.

Monardes, N. 1925. *Joyfull Newes out of the Newe Founde Worlde*, trans. J. Frampton. London: Constable and Co. (orig. 1577).

Moreno de los Arcos, R. 1967. Los cinco soles cosmogónicos. *Estudios de Cultura Nahuatl* 7: 183–210.

Moriarty, J. R. 1974. Aztec jade. *Katunob* 10 (4): 9–14.

Morton, J. F. 1977. *Major Medicinal Plants*. Springfield, IL: Charles C. Thomas.

———. 1981. *Atlas of Medicinal Plants of Middle America*. Springfield, IL: Charles C. Thomas.

Morton, P. 1987. Psychology and the immune system. *New Scientist* (9 April): 46–50.

Moss, L. 1983. Personal communication.

Moss, L. and S. C. Cappannari. 1976. Mal'occhio, ayinha ra, oculus fascinus, judenblick: The evil eye hovers above. In *The Evil Eye*, ed. C. Maloney, 1–15. New York: Columbia Univ. Press.

Motolinía, Fr. T. de Benavente o. 1971. *Memoriales o Libro de las Cosas de la Nueva España y de los Naturales de Ella*, ed. E. O'Gorman. México: UNAM (orig. 1541).

Naeve, R. L. 1979. Coitus and associated amniotic fluid infections. *New England Journal of Medicine* 301: 1198–1200.

Nakamura, H. 1982. *Spirulina: Food for a Hungry World*. Boulder Creek, CA: Univ. of the Trees Press.

National Research Council. 1984. *Amaranth. Modern Prospects for an Ancient Crop*. Washington, D.C.: National Academy Press.

Navar, L. G. 1987. Personal communication.

Neher, R. T. 1968. The ethnobotany of Tagetes. *Economic Botany* 22: 317–327.

Newman, M. T. 1976. Aboriginal New World epidemiology and medical care, and the impact of Old World disease imports. *American Journal of Physical Anthropology* 45: 667–672.

Nicholson, H. B. 1971. Religion in pre-Hispanic central Mexico. In *Handbook of Middle American Indians*, ed. G. E. Eckholm and I. Bernal, vol. 10, 395–446. Austin: Univ. of Texas Press.

Office of Technology Assessment. 1978. *Assessing the Efficacy and Safety of Medical Technologies*. Washington, D.C.: U.S. Government Printing Office.

Orellana, S. L. 1987. *Indian Medicine in Highland Guatemala*. Albuquerque: Univ. of New Mexico Press.

Ortiz de Montellano, B. R. 1974/75. Aztec medicine: Empirical drug use. *Etnomedizin* 3: 249–271.

————. 1975. Empirical Aztec medicine. *Science* 188: 215–220.

————. 1976. Curanderos: Spanish shamans and Aztec scientists? *Grito del Sol* 1 (2): 21–28.

————. 1978. Aztec cannibalism: An ecological necessity? *Science* 200: 611–617.

————. 1979a. Mesoamerican calendar: Philosophy and computations. *Grito del Sol* 4 (2): 49–73.

————. 1979b. The rational causes of illness among the Aztecs. *Actes du XLIIe Congrès International des Américanistes* (Paris 1976) 6: 287–299.

————. 1980a. Las yerbas de Tlaloc. *Estudios de Cultura Nahuatl* 14: 287–314.

————. 1980b. Minorities and the medical professional monopoly. *Grito del Sol* 5 (2): 25–63.

————. 1981. Entheogens: The interaction of biology and culture. *Reviews in Anthropology* 8: 339–363.

————. 1984. El conocimiento de la naturaleza entre los mexicas. Taxonomia. In *Historia General de la Medicina en México*, vol. 1, ed. A López Austin and C. Viesca Treviño, 115–132. Mexico: UNAM–Academia Nacional de la Medicina.

————. 1986. Aztec sources of some mexican folk medicine. In *Folk Medicine. The Art and the Science*, ed. R. P. Steiner, 1–22. Washington, D.C.: American Chemical Society.

————. 1987. *Caída de mollera*: Aztec sources for a Mesoamerican disease of alleged Spanish origin. *Ethnohistory* 34: 381–399.

————. 1989a. Ghosts of the imagination: John Bierhorst's translation of "Cantares Mexicanos." *New Scholar* 11: in press.

————. 1989b. Mesoamerican religious tradition and medicine. In *Caring and Curing in the World's Religious Traditions*, ed. L. E. Sullivan, 359–394. New York: Macmillan.

Ortiz de Montellano, B. R. and C. H. Browner. 1985. Chemical bases for medicinal plant uses in Oaxaca, Mexico. *Journal of Ethnopharmacology* 13: 57–88.

Page, E. W., C. A. Villee, and D. B. Villee. 1976. *Human Reproduction*, 2d ed. Philadelphia: W. B. Saunders.

Palerm, A. and E. Wolf. 1972. *Agricultura y Civilización en Mesoamérica*. México: SEP-Setentas.

Pan American Health Organization. 1986. *Health Conditions in the Americas 1981–1984*. Washington, D.C.: Panamerican Health Organization.

Parsons, J. R. 1976. The role of chinampa agriculture in the food supply of Aztec Tenochtitlan. In *Cultural Change and Continuity: Essays in*

Honor of James Bennett Griffin, ed. C. Cleland, 233–262. New York: Academic Press.

Parsons, J. R., M. H. Parsons, V. Popper, and M. Taft. 1985. Chinampa agriculture and Aztec urbanization in the Valley of Mexico. In *Prehistoric Intensive Agriculture in the Tropics*, ed. I. S. Farrington, 49–96. Oxford: B.A.R. International Series 232.

Patrick, L. L. 1977. A cultural geography of the use of seasonally dry sloping terrain: The metepantli crop terraces of Central Mexico. Ph. D. dissertation. Univ. of Pittsburgh.

Payne, P. R. 1975. Safe protein-calorie ratios in diets. The relative importance of protein and energy intakes as causal factors in malnutrition. *American Journal of Clinical Nutrition* 28: 281–286.

Perez-Cirera, R. 1944. Contribucíon al estudio farmacológico de la *Commelina pallida*. I. Accíon coagulante y vasoconstrictora. *Gaceta Médica de México* 74: 140–145.

Phelan, J. L. 1972. *El Reino Milenario de los Franciscanos en el Nuevo Mundo*. México: UNAM.

Pliny (Gaius Plinius Secundus). 1949–1962. *Natural History*, trans. W.H.S. Jones, 10 vols. Cambridge: Harvard Univ. Press (orig. A.D. 77).

Pomar, J. B. de. 1964. Relacíon. In *Poesía Nahuatl*, ed. A. M. Garibay, vol. 1, 149–219. México: UNAM (orig. 1582).

Ponce de León, P. 1973. Tratado de los dioses y ritos de la gentilidad. In *Teogonía e Historia de los Mexicanos: Tres Opúsculos del Siglo XVI*, ed. A. M. Garibay, 121–132. México: Porrúa (orig. 1569).

Price, B. J. 1978. Demystification, enriddlement, and Aztec cannibalism: A materialist rejoinder to Harner. *American Ethnologist* 5: 98–115.

Quezada, N. 1975a. Métodos anticonceptivos y abortivos tradicionales. *Anales de Antropología* 12: 223–242.

———. 1975b. *Amor y Magia Amorosa entre los Aztecas*. México: UNAM.

Quiroz-Rodiles, A. 1945. Breve historia de la obstetricia en Mexico. *Obstetricia y Ginecología Latino-Americanas* 3: 77–92.

Ramos de Elurdoy, J. 1982. *Los Insectos como Fuente de Proteinas en el Futuro*. México: Editorial Limusa.

Ransford, O. 1983. *Bid the Sickness Cease*. London: John Murray.

Read, K. A. 1986. The fleeting moment: Cosmogony, eschatology, and ethics in Aztec religion and society. *The Journal of Religious Ethics* 14: 113–138.

———. 1987. Negotiating the familiar and the strange in Aztec ethics. *The Journal of Religious Ethics* 15: 2–13.

Recinos, A., trans., and D. Goetz and S. G. Morley, eds. 1950. *Popol Vuh*. Norman: Univ. of Oklahoma Press.

Redfield, R. and A. Villa Rojas. 1964. *Chan Kom. A Maya Village*, 2d ed. Chicago: Univ. of Chicago Press.

Reichel Dolmatoff, G. 1978. *Beyond the Milky Way*. Los Angeles: Univ. of California Press.

Rennie, M. J. and R. Harrison. 1984. Effects of injury, disease, and malnutrition on protein metabolism in man. *Lancet i* (February 11): 323–325.

Ricard, R. 1966. *The Spiritual Conquest of Mexico*, trans. L. B. Simpson. Berkeley: Univ. of California Press.

Riley, V. 1981. Psychoneuroendocrine influences on immunocompetence and neoplasia. *Science* 212: 1100–1109.

Risse, G. 1984. Personal communication.

Robertson, D. 1959. *Mexican Manuscript Painting of the Early Colonial Period*. New Haven: Yale Univ. Press.

Robicsek, F. and D. M. Hales. 1984. Maya heart sacrifice: Cultural perspectives and surgical technique. In *Ritual Human Sacrifice in Mesoamerica*, ed. E. H. Boone, 49–90. Washington, D.C.: Dumbarton Oaks.

Rozanski, A. et al. 1988. Mental stress and the induction of silent myocardial ischemia in patients with coronary artery disease. *New England Journal of Medicine* 318: 1005–1012.

Rubel, A. J. 1960. Disease in Mexican-American culture. *American Anthropologist* 62: 795–814.

———. 1964. The epidemiology of a folk illness: Susto in Hispanic America. *Ethnology* 3: 268–283.

Rubel, A. J., C. W. O'Neill, and R. Collado-Ardón. 1984. *Susto, A Folk Illness*. Berkeley: Univ. of California Press.

Ruiz de Alarcón, A. 1953. *Tratado de las Idolatrías, Supersticiones, Dioses, Ritos, Hechizerías y otras Costumbres Gentílicas de las Razas Aborígenes*. México: Ediciones Fuente Cultural (orig. 1629).

———. 1982. *Aztec Sorcerers in Seventeenth Century Mexico. The Treatise on Superstitions by Hernando Ruiz de Alarcón*, ed. and trans. M. D. Coe and C. Whittaker. Albany: Inst. Mesoamerican Studies, SUNY (orig. 1629).

———. 1984. *Treatise on the Heathen Superstitions*, ed. and trans. J. R. Andrews and R. Hassig. Norman: Univ. of Oklahoma Press (orig. 1629).

Sahagún, Fr. B. de. 1906. *Códice Matritense del Real Palacio*, ed. F. del Paso y Troncoso. Madrid: Hauser y Manet (facsim., orig. 1560–1565).

———. 1950–1969. *Florentine Codex. General History of the Things of New Spain*, ed. and trans. C. E. Dibble and A. J. O. Anderson, 12 books. Salt Lake City: Univ. of Utah Press (orig. 1577).

———. 1956. *Historia General de las Cosas de la Nueva España*, ed. A. M. Garibay, 4 vols. México: Porrúa (orig. 1793).

———. 1986. *Coloquios y Doctrina Christiana*, ed. and trans. M. León Portilla. México: UNAM (orig. 1524).

Sahlins, M. 1978. Culture as protein and profit. *New York Review of Books* November 23: 45–53.

Sanders, W. T. 1976a. The population of the Central Mexico symbiotic region, the Basin of Mexico, and the Teotihuacan Valley in the Sixteenth Century. In *The Native Population of the Americas in 1492*, ed. W. M. Denevan, 85–155. Madison: Univ. of Wisconsin Press.

———. 1976b. The Agricultural History of the Basin of Mexico. In *The Val-*

ley of Mexico. Studies in pre-Hispanic Ecology and Society, ed. E. R. Wolf, 101–159. Albuquerque: Univ. of New Mexico Press.

Sanders, W. T., A. Kovar, T. Charlton, and R. A. Diehl. 1970. *The Natural Environment, Contemporary Occupation and Sixteenth Century Population of the Valley (Teotihuacan Valley Project Final Report)*. University Park: Pennsylvania State Univ.

Sanders, W. T., J. R. Parsons, and R. S. Santley. 1979. *The Basin of Mexico. Ecological Processes in the Evolution of a Civilization*. New York: Academic Press.

Santillán, C. 1982. Mass production of *Spirulina. Experientia* 38: 40–43.

Santley, R. S. and E. K. Rose. 1979. Diet, nutrition and population dynamics in the Basin of Mexico. *World Archeology* 1 (2): 185–207.

Scheffler, L. 1977. Medicina folk y cambio social en un pueblo nahuatl del Valle de Tlaxcala. *Boletin del Departamento de Investigacion de las Tradiciones Populares (Mexico, Secretaria de Educacion Publica)* 4: 83–107.

Schele, L. and M. L. Miller. 1986. *The Blood of Kings*. New York: George Braziller.

Scheper-Hughes, N. and D. Stewart. 1983. Curanderismo in Taos County, New Mexico—A possible case of anthropological romanticism. *Western Journal of Medicine* 139: 875–884.

Schleifer, S. J., S. E. Keller, M. Camerino, J. C. Thornton, and M. Stein. 1983. Suppression of lymphocyte stimulation following bereavement. *Journal American Medical Association* 250: 374–377.

Schultes, R. E. and A. Hofmann. 1979. *Plants of the Gods*. New York: McGraw Hill.

———. 1980. *The Botany and Chemistry of Hallucinogens*, 2d ed. Springfield, IL: Charles C. Thomas.

Seler, E. 1900–1901. *The Tonalamatl of the Aubin Collection*. London: Hazell, Watson and Viney.

———. 1960–1969. Die religiosen gesange der alten Mexikaner. In *Gessamelte Abhandlungen zur Amerikanischen Sprach-und Altertumskunde*, vol. 7, 961–1107. Graz, Austria: Akademische Druck.

Sentíes, G. and R. Amayo. 1964. Efecto del cihuapatli sobre el útero humano grávido. *Gaceta Médica Mexicana* 4: 343–350.

Sigerist, H. C. 1967. *A History of Primitive and Archaic Medicine*. New York: Oxford Univ. Press.

Signorini, I. 1982. Patterns of fright: Multiple concepts of susto in a Nahua-Ladino community of the Sierra de Puebla (Mexico). *Ethnology* 21: 313–323.

Simmons, O. 1955. Popular and modern medicine in mestizo communities of coastal Peru and Chile. *Journal of American Folklore* 68: 57–71.

Singer, C. 1957. *A Short History of Anatomy and Physiology from the Greeks to Harvey*. New York: Dover.

Sklar, L. S. and H. Anisman. 1979. Stress and coping factors influence tumor growth. *Science* 205: 513–515.

Somolinos d'Ardois, G. 1960. Vida y obra de Francisco Hernández. In F. Hernández, *Obras Completas*, vol. 1, 97–440. México: UNAM.

———. 1964. Estudio histórico. In M. de la Cruz, *Libellus de Medicinalibus Indorum Herbis*, ed. E. C. del Pozo, 301–327. México: IMSS.

Soustelle, J. 1970. *Daily Life of the Aztecs*. Stanford: Stanford Univ. Press.

Southam, L. et al. 1983. The zoapatle IV—Toxicological and chemical studies. *Contraception* 27: 255–265.

Spiro, H. M. 1986. *Doctors, Patients and Placebos*. New Haven: Yale Univ. Press.

Spitz, W. 1977. Wayne County Medical Examiner, personal communication.

Spoerke, D. G. 1980. *Herbal Medications*. Santa Barbara, CA: Woodbridge Press.

Steinegger, E. and R. Haensel. 1963. *Lehrbuch der Allgemeine Pharmakognosie*. Berlin: Springer Verlag.

Stewart, P., E. Hilton, and A. A. Calder. 1983. A randomized trial to evaluate the use of a birth chair for delivery. *Lancet i* (June 11): 1296–1298.

Stewart, T. D. 1960. A physical anthropologist's view of the peopling of the New World. *Southwest Journal of Anthropology* 16: 259–273.

Sticher, O. 1977. Plant mono-, di-, and sesquiterpeneoids with pharmaceutical or therapeutic activity. In *New Natural Products and Plant Drugs with Pharmacological, Biological, or Therapeutical Activity*, ed. H. Wagner and P. Wolff, 137–176. New York: Springer Verlag.

Taylor, R. L. 1975. *Butterflies in My Stomach or Insects in Human Nutrition*. Santa Barbara, CA: Woodbridge Press.

Temkin, O. 1956. *Soranus Gynecology*. Baltimore: Johns Hopkins Univ. Press.

Terman, G. W., Y. Shavit, J. W. Lewis, J. T. Cannon, and J. C. Liebeskind. 1984. Intrinsic mechanisms of pain inhibition: Activation by stress. *Science* 226: 1270–1277.

Thompson, J.E.S. 1966. *The Rise and Fall of Maya Civilization*. Norman: Univ. of Oklahoma Press.

Thompson, W.A.R. 1978. *Medicines from the Earth*. New York: McGraw Hill.

Torquemada, Fr. J. de. 1975–1983. *Monarquía Indiana*, ed. M. León Portilla, 7 vols. México: UNAM (orig. 1615).

Torrey, E. F. 1986. *Witchdoctors and Psychiatrists*. New York: Harper & Row.

Trease, G. E. and W. C. Evans. 1978. *Pharmacognosy*, 11th ed. London: Balliere Tyndall.

Trotter, R. T. 1982. Susto: The context of a community morbidity pattern. *Ethnology* 21: 215–226.

———. 1985. Folk medicine in the Southwest. *Postgraduate Medicine* 78: 167–179.

Trotter, R. T., O. C. Garza, and M. Garza. n.d. Caída de mollera: Health risks from a Mexican-American folk illness. Typescript.

Trouillet, J. L. et al. 1985. Use of granulated sugar in treatment of open mediastinitis after cardiac surgery. *Lancet ii* (July 27): 180–184.

Truman, K. A. n.d. Colonial domination and capitalist production in Mexico and Guatemala. Typescript.

Tschesche, R. and G. Wulff. 1972. Chemie und biologie der saponine. *Fortschritte des Chemie Organischer Naturstoffe* 30: 461–604.

Turner, P., ed. 1964. *Selections from the "Natural History" of C. Plinius Secundus*, trans. Philemon Holland. New York: McGraw Hill (orig. 1642).

Tyler, V. E. 1987. *The New Honest Herbal.* Philadelphia: George F. Stickley Co.

Vaillant, G. C. 1966. *Aztecs of Mexico.* New York: Penguin Books.

Valdizán, H. and A. Maldonado. 1922. *La Medicina Popular Peruana*, 3 vols. Lima: Imprenta Torrez Aguirre.

Vargas, L. A. and E. Matos Moctezuma. 1973. El embarazo y el parto en el México prehispanico. *Anales de Antropología* 10: 301–310.

Vayda, A. P. 1970. On the nutritional value of cannibalism. *American Anthropologist* 72: 1462–1463.

Viesca Treviño, C. 1981. Personal communication.

——. 1982. Hambruna y epidemia en Anahuac en la época de Moctezuma Ilhuicamina. In *Ensayos sobre la Historia de las Epidemias en México*, ed. E. Florescano and E. Malvido, vol. 1, 157–177. México: IMSS.

——. 1984a. Epidemiología entre los Mexicas. In *Historia General de la Medicina en México*, ed. A. López Austin and C. Viesca Treviño, vol. 1, 171–187. México: UNAM.

——. 1984b. Prevencíon y terapeuticas Mexica. In *Historia General de la Medicina en Mexico*, ed. A. López Austin and C. Viesca Treviño, vol. 1, 201–216. México: UNAM–Academia Nacional de Medicina.

——. 1984c. El médico Mexica. In *Historia General de la Medicina en Mexico*, ed. A. López Austin and C. Viesca Treviño, vol. 1, 217–230. México: UNAM–Academia Nacional de Medicina.

Viesca Treviño, C. and I. de la Peña. 1974. La magia en el Códice Badiano. *Estudios de Cultura Nahuatl* 11: 267–301.

——. 1979. Las crisis convulsivas en la medicina Nahuatl. *Anales de Antropología* 14: 487–495.

Vogel, V. J. 1970. *American Indian Medicine.* New York: Ballantine Books.

von Reis, S. and F. J. Lipp. 1982. *New Plant Sources for Drugs and Foods.* Cambridge: Harvard Univ. Press.

von Reis Altschul, S. 1973. *Drugs and Food from Little-Known Plants.* Cambridge: Harvard Univ. Press.

Walens, S. and R. Wagner. 1971. Pigs, protein and people-eaters. *American Anthropologist* 73: 269–270.

Wasson, R. G. 1980. *The Wondrous Mushroom. Mycolatry in Mesoamerica.* New York: McGraw Hill.

Waterlow, J. C. and P. R. Payne. 1975. The protein gap. *Nature* 258: 113–117.

Waterlow, J. C., M. Golden, and D. Perou. 1977. Measurement of rates of protein turnover, synthesis, and breakdown in man and the effects of nutritional status and surgical injury. *American Journal of Clinical Nutrition* 30: 1333–1339.

Weil, A. 1983. *Health and Healing*. Boston: Houghton Mifflin.

Whitmore, T. and B. L. Turner. n.d. Population reconstruction of the Basin of Mexico: 1150 B.C. to present (mimeo.).

Willer, J. C., H. Deher, and J. Cambier. 1981. Stress-induced analgesia in humans: Endogenous opiods and naloxone-reversible depression of pain reflexes. *Science* 226: 1270–1277.

Witherspoon, G. 1977. *Language and Art in the Navajo Universe*. Ann Arbor: Univ. of Michigan Press.

Wood, C. S. 1975. New evidence for a late introduction of malaria into the New World. *Current Anthropology* 16: 93–104.

———. 1979. *Human Sickness and Health: A Biocultural View*. Palo Alto: Mayfield.

Ximénez, F. 1888. *Cuatro Libros de la Naturaleza y Virtudes Medicinales de las Plantas y Animales de la Nueva España*, ed. N. León. Morelia, México: Escuela de Artes (orig. 1615).

Young, J. C. and L. Young-Garro. 1982. Variation in the choice of treatment in two medical communities. *Social Science & Medicine* 16: 1453–1465.

Young, P. 1981. Aztec surgical blade. Medicine and archeology wedded. *Detroit Free Press* (February 9): 3A.

Ysunza-Ogazón, A. 1976. Estudios bio-antropologicos del tratamiento del susto. In *Estudios sobre Etnobotánica y Antropología Médica*, ed. C. Viesca Treviño, 59–73. Mexico: IMEPLAM.

Zelitch, I. 1979. Photosynthesis and plant productivity. *Chemistry and Engineering News* (February 5): 28–48.

Index